A FISH COME TRUE

A FISH COME TRUE

Fables, Farces, and Fantasies for the Hopeful Angler

PAUL SCHULLERY

STACKPOLE
BOOKS

Guilford, Connecticut

STACKPOLE
   BOOKS
Published by Stackpole Books
An imprint of The Rowman & Littlefield Publishing Group, Inc.
4501 Forbes Blvd., Ste. 200
Lanham, MD 20706
www.rowman.com

Distributed by NATIONAL BOOK NETWORK

British Library Cataloguing in Publication Information available

Library of Congress Cataloging-in-Publication Data

Names: Schullery, Paul, author.
Title: A fish come true : fables, farces, and fantasies for the hopeful
   angler / Paul Schullery.
Description: Guilford, Connecticut : Stackpole, [2019] |
Identifiers: LCCN 2019010675 (print) | LCCN 2019011480 (ebook) | ISBN
   9780811768634 (electronic) | ISBN 9780811738651 (cloth : alk. paper)
Subjects: LCSH: Fishing—Anecdotes.
Classification: LCC SH441 (ebook) | LCC SH441 .S3827 2019 (print) | DDC
   639.2—dc23
LC record available at https://lccn.loc.gov/2019010675

♾️™ The paper used in this publication meets the minimum requirements of American National Standard for Information Sciences—Permanence of Paper for Printed Library Materials, ANSI/ NISO Z39.48-1992.

For absent friends

Craig Woods
Richard Kress
David Detweiler

The charm of fishing is that it is the pursuit of what is elusive but attainable, a perpetual series of occasions for hope.
—JOHN BUCHAN

CONTENTS

Preface
What If?

OVER THE YEARS THAT I HAVE OCCASIONALLY WORKED ON THE stories that became this book I often considered calling it *A Fisherman's What-if Tales*, because in their very different ways that's what all these stories turned out to be. What if someone discovered a fly that worked on every cast? What if we could fish any where, any time, in the distant past? What if we could explore the fishing on a different planet? What if our sport's leading thinkers suddenly decided that an infamous "trash fish" was really cool and a worthy sporting trophy after all? What if a great American artist and a pioneering American fly theorist met to share their most penetrating insights about how their respective passions could serve each other? What if in some unfathomable way having nothing to do with our skill as anglers, some mysterious personal worthiness allows us to catch the fish of a lifetime? What if "Me and Joe," those immortal anglers of troubled literary tradition, weren't quite the innocent simpletons that so many later commentators have presumed them to be?

I imagine most of us find it satisfying, even unavoidable, to exercise our wildest hopes about fishing. It's yet another way we can enjoy this wonderful sport even when we're not out there blundering around on our favorite stream. And in that spirit I offer this book, which explores at least a few of the hopes, fears, laughs, surprises, and ironies that might arise from the fulfillment of our angling dreams.

Having only a passing interest in considering such possibilities in any disciplined or lofty way—what *is* a happy ending, anyway?—I have instead gone for atmosphere, provocation, and whimsy. And although, as I read these stories now it looks to me like whimsy mostly wins out, there are spots in some of them that do aim for something a little more substantial. When the final story in the bunch, "Shupton's Fancy," was originally published, the reviewers and blurbers were gratifyingly generous about its depths; someone even used the wondrous term "literature." I couldn't ask for more than that.

So I will let this preface go at that except to offer some thanks that seem to me too important merely to bury in the fine print at the back of the book.

The stories were mostly my idea. The most important exception involves chapter 5, which I am pleased to say was inspired by an unforgettable science fiction story, "A Sound of Thunder," written by the great Ray Bradbury and first published in *Collier's* magazine in 1952 (and republished often thereafter). Readers of Bradbury's work will recognize the common theme of time-traveling sportsmen; his were hunters, mine are anglers.

Another exception is chapter 8, about an off-in-the-distant-future fly-fishing expedition to planet Kepler-22b. A few years ago, when scientists declared that Kepler-22b, an extrasolar planet about six hundred light-years from Earth, might be covered in water, Marshall Cutchin, creator and editor of the splendid Mid current.com, sent out an invitation to write a story about what it might be like to fish there. I couldn't resist, and the result is the pair of divergent press releases relating to that fishing trip.

Otherwise, all I can do here is thank you for your kind attention to my books and wish you good fishing wherever your dreams may lead you.

Paul Schullery
In the Gallatin Valley, Montana
2019

Second Season

THE FAMILY STATION WAGON BUMPED ALONG THE CANYON ROAD. The people weren't lost, they knew where they were and where they were heading, but this wasn't the road they wanted. It wasn't a good road, sometimes little better than packed dirt, and it followed the high contours of the canyon wall, so it was full of sharp turns. The husband was peeved about the lost time and wondered how they'd missed the turn to the interstate. His wife was studying the map, but now she lowered it to her lap and looked out the window. It was a rather monotonous canyon, she thought, each bend like the last, and most of the steep slopes had no trees—just a lot of scrubby stuff, almost like a desert. And the river didn't appeal to her. As they crossed a bridge over a small creek she looked down at the river. There were two men standing in the water, fishing.

He had been on the river for three days now and had taken no fish. There was a strike early the second day, just as his line was finishing its swing. He didn't see the fish, just felt its surge for an instant and then it was just the river again and his fly ticking on a rock in the shallows as he retrieved the line. He smiled, feeling good that he'd even detected the fish's take. Had he been a little less observant, a little less tuned to the current, he might not even have noticed that the line stopped, ever so slightly offstream, only a couple degrees from hanging straight downstream, still in the current. Even before the line could belly, he knew that a fish had taken the fly. Though he was puzzled when the hook came free, he wasn't especially disappointed. The fish weren't taking. Even the guide boats that rumbled past periodically were hard pressed for action. It was a matter of waiting, for a change of weather or light or whatever the river needed to move its fish.

He wasn't worried about having the right tackle, either. He smiled again, remembering how he'd struggled the year before with the bewildering array of flies, trying to carry a representative selection on his first trip; and how he'd agonized over rods and lines, not knowing that most of his choices would work just fine. He'd learned what all his puzzling had been worth when he got to the river. He met other fishermen and talked to them, but he had the feeling that they knew no more than he did. Then, the day he caught his first small fish he met three veteran steelhead fishermen and more or less attached himself to them. Watching them, listening to them, and finally fishing with them, he had learned the essentials.

The flies, it seemed, were least important. Their flies were lightly dressed, no patterns he recognized from the books he'd read, more in the spirit of low-water Atlantic salmon flies than in the bulky, garish style of the book-taught patterns he'd so dutifully tied up. But he soon learned that they used these lighter patterns primarily because they enjoyed them most, and that his lumpy chenille patterns were probably just as likely to work. No one really knew.

Their fly rods were just as improbable. One man used a light trout outfit, another an outfit that seemed better suited to tarpon than steelhead. All three of them could cast with phenomenal power, though no two had the same style. They fished differently, but they were a team, spacing themselves evenly as they worked their flies through a run until all were at the lower end.

From them, he learned what was important: the presentation, the greased-line technique that somehow brought fish up from the bottom, fish that ignored weighted flies put almost in their mouths but moved inexplicably through several vertical feet of current to take the barely submerged, sparsely dressed fly on the end of a properly mended line. "They'll come up for it" they told him.

So he joined them, impatient with his awkward casts and looping line that seemed to droop and hang where theirs whipped straight. He swore then that when he came back he would be better at all this.

And he was better this time. It was the third fishless day, but he was confident in the moves of the fishing. His timing was sometimes off on the hauls, but he could throw the long line and make the big mend when it broke in the current.

He was fishing slowly, covering the water with geometric precision as he searched the likely spots. He remembered the jolt he'd gotten hearing the old-timers discussing the water they'd just fished. They spoke of river-bottom features that he, on the river several days before them, had hardly noticed. He began to search the water as they talked, and slowly things came into focus— submerged rocks, shallows, slicks, pockets, all the places he had learned to watch for in trout fishing but had somehow ignored here on the river. He was overcome and buffaloed by the newness of steelhead water. He had stopped thinking. He remembered that now as he worked the good water with closely spaced casts.

He stopped when the sun hit the water. A lot of people did, and he wasn't convinced the fish took it that seriously, but it was a nice excuse to take a break. The only other fisherman around, upstream at the head of the next run, was also coming out and they walked toward each other as if it was planned. He'd seen the fellow the day before, reading him as an old local, probably retired. He was tall and heavy, but never waded above his knees and had a stiff mechanical cast like someone whose elbow was frozen at a ninety-degree bend.

They sat on some rocks and talked about the fishing, then the old man, sensing a good audience, opened up. "You know, my wife, she used to accuse me of coming up here just to get away from her. Said I didn't like to fish, just wanted to get out of the house, like she thought I could only stand so much of her. Then she died, been three years this October, and I still come up here every chance I get. Guess that sort of proves her wrong, though it's kind of late. She never did figure it out, you know? People like different things. Take the gulls. To me they're just like rats at a dump, picking up trash they find. They'll eat the eyes out of your fish, you leave them on the ground; and they're always screeching around, acting like rats. But you take someone from inland, why, they see one of those gulls and get all excited, feed it bread and try to get it to land on their shoulder or something. There's no explaining it."

"It's like that with fishing, too." The old man was rolling now, and he listened to him while looking across the rocks to the river, now half-shaded, half-sunlit. "My neighbors used to try to talk to me about it, then they caught on that I didn't care what they thought, that I wasn't fishing to feed myself, that it wasn't something between friends, you know? I mean they thought I was out here for the food, or the exercise, or to get away from my wife, for God's sake!"

All at once the old man seemed done talking and got up to go, not even saying good-bye. But the river still looked good so he decided to fish some more, the shaded water at the tail of the run.

As he waded in he heard the clatter of rocks as the old man scrambled up the slope to his truck. He wondered how many times the guy had told that story, made that little speech, and wondered where he'd heard that part about the wife dying before. It didn't sound quite original. Then he wondered if he too had an *apologia*, a canned justification for social occasions that demanded a defense of such a seemingly frivolous pastime.

The two women he was intermittently seeing both tolerated the fishing well, but he wasn't sure why. Sometimes he worried that they thought he still had a lot of the little boy in him. That irritated him; he hated being patronized. Either of them was quite willing to go with him when he went fishing, something that made him uncomfortable because neither would fish and he always quit early because he thought that they must be bored sitting on the bank reading or just watching him. At least, he thought, at least I'm not being accused of doing it just to get away.

The shaded band of river was along the far shore, and the casts were demanding, making him concentrate on his rhythm and taking him from other thoughts for a moment. Then he missed the timing and felt the fly click hard on a rock behind him. He swore, more from habit than from real anger, and began to draw the line in to examine the hook. He recalled that he'd been thinking about the girls, but he consciously determined to think about something else.

The bad cast reminded him of his first trip with the big rod, how he would work out more line than he could handle and actually shout as he pulled the final backcast forward, as though the vocal energy would boost his arm power and help shoot the last bit of line. It was fun with the river so big and loud that he could loose a yell across it now and then.

He was nearing the tail of the shaded water, where the run began to break, showing white here and there before it turned into a long shallow rapid. He was casting almost directly across the current, mending hard, and as the line reached the corner—the middle of its quarter swing—the fish hit.

It was not the dull munching bump of a stale fish, taking out of convenience, but the jarring attack, always surprising, always thrilling, of a fresh steelhead. The initial downstream run, down through the rapid and into the next long run, took him fifty yards into his backing. He didn't get a good look at the fish because he was running too, splashing through the shallows and scrambling over the rocks trying to get below the fish. But he saw enough to know it was a large fish, large even for the river. For an instant, running and winding line and trying to watch the river where the line disappeared into it, he wished the old man had stayed.

The fish was holding in midstream about halfway down this run, and he regained nearly all the backing before it ran again downstream. He noticed the chatter of his reel this time, and felt a bit hopeless about ever getting below a fish this strong. He remembered a man he'd seen the year before, foul-hooked to a wildly running ten-pounder, and the trail of fly boxes and other stuff the man spilled from his vest along a quarter-mile of stream as he chased the fish.

The river was featureless below the run, with no obvious obstructions, no distinct channels or real pools, just wide and deep. He hoped the fish would feel fairly secure in such big water and let him move around it a little more. It held its position after the second run, again in midstream, while he cranked in the backing and stumbled to a point on the shore even with the fish. He was applying very little pressure, risking a thrown hook for a chance to get downstream of the fish and make it run against the current. He remembered seeing a man throw rocks below a fish to get it to break upstream, remembered doing the same thing himself with a big old brown trout he'd once hooked at home on hopelessly light tackle. This fish was deep, though, and he didn't really think it would work in such big water, and he'd always kind of thought that throwing rocks was probably bad manners, or unsporting, or something. Still, if he decided to do it, he was glad the old man wasn't there to watch.

He'd gotten even with smaller steelhead before, even so that by holding his rod away from his body, pointed downstream, the rod tip was actually below the fish, but he'd never gotten well below one. This one let him get even three times, each time moving downstream again, still very strong. Chases like this were full of risks, of flies working loose, of submerged logs or rocks, of jammed reels, and he wished he could stop it. He came even with the fish again and without giving it any thought raised the rod to reset the hook, just as a precaution. The sudden stab of pressure, or whatever passes for pain in the fish's physiology, sent the fish off again, only this time upstream in a long run climaxed by a series of cartwheeling leaps so close together that he couldn't imagine where the fish found the momentum for each subsequent jump. He was at last, however, well downstream of the fish, and it no longer had the current working for it.

He was watching those last jumps carefully and saw that it was an exceptional fish. He began to wonder what he should do with it if he landed it. It would be great to show it to someone, but he didn't have his camera along. He might have to kill it if he wanted anyone to believe he'd caught such a big one. He wondered what the record for the river was.

The fish stopped moving around and began to make strong yanks on the line from one spot in midstream. He felt the tugs and twists as it tried to shake free. He increased the pressure a little more, and the fish moved upstream and closer to the near shore. It was now at the head of the run, almost to the rapids, and when he tried to move up on it, it bolted, first into the rapid itself and then back downstream to a boulder that barely broke the surface. He saw it flash and roll then, and he thought it was going to jump but suddenly it twisted down behind the rock. Standing about seventy feet from the rock, he decided to try to goad the fish into action again. He pushed the rod butt forward, but met unyielding resistance. He swore as he realized that the line was hung up on the rock above the fish.

In an effort to free the line, he hurried up shore of the rock, reaching as far out and up as he could, but the line wouldn't give. The fish appeared to be about six feet downstream from the rock, so he couldn't tell if the snag was on the line or the leader. Probably the connecting loop, he thought, or the knot that held the leader butt to the line. When he stood at the edge of the water directly across from the fish, it was no more than thirty feet away. He could see it moving smoothly in the current, very big, frightfully big, and wondered if he could wade to it. He figured the boulder was in three or four feet of water, and current was slow here, flattened out after the rapid. He was not a confident wader for all his size, but it was worth a try.

He planned to bring the fish back by hand, so he took his rod upstream a few yards and wedged the handle between some rocks so that most of the line was out of the water. He stepped into the water and quickly covered the first twenty feet. The water came to his thighs and was pushing firmly against him, whirling fine gravel from under his feet. The fish seemed to be ignoring him. He could see that the water got deeper faster between him and the rock, and that opaque dark green color of real depth just beyond the rock. He could see the line too, tightly wrapped under a ledge along the side of the rock about a foot below the surface.

When the water reached his waist he was still five feet from the fish. He'd started walking toward the rock but the force of the current had put a curve in his course and he was going to reach the fish at its tail if he was lucky. His footing was poor, and he began to feel the first touches of buoyancy that come from wading deep. He was giddy, reminded of the panic he felt once on a big Michigan river when he was almost washed into a deep pool.

Barely lifting his feet, he inched across the bottom, correcting his course and angling upstream. A high wave slopped over his waders and he realized he was crouching. He straightened slightly, reached out, and caught the leader just above the fish's mouth, He expected a furious struggle but the fish didn't react as he pulled it to him and slipped two fingers deep into its gill. It occurred to him that he had decided to kill it, but he couldn't remember when. As he drew the fish up he could see that it had been bleeding heavily from the gills, probably the result of some injury during its struggle, maybe something to do with the boulder it was hanging from. It behaved now as if it was almost dead.

Wading directly upstream up to the boulder to free the line was out of the question; he was already tired from the strain. So he deepened his hold in the fish's gill and clipped the leader. He would have to take his chances with freeing the line when he got back to shore. He was glad he hadn't brought the rod along.

Going back was easier. There was one sickening slip just as he started and he almost went over, but his preventive lurch landed his feet on solid bottom and he was able to pick his way at a downstream angle until he was in the shallows.

He slogged up onto the shore, lay the fish down, and suddenly felt very cold. He'd shipped much more water than he remembered and his shirt and pants were quite wet. He walked up to where the rod was propped, picked up the line, wrapped it twice around his sleeve, and gave it a couple good yanks before it snapped back from the boulder. He could see from the break that the snag had been at the knot.

The sun was high now and the whole river was out of shade. He peeled his waders down to his boots and sat down to let the sun dry him out, but after a moment he thought of the fish, so he got up and waddled over to it. The fly was still embedded in the back of the jaw, and a little blood had run from the gills onto the dark rocks. Later, he thought, I'll take it and get it weighed somewhere. He moved it to where he could look at it as he rested in the sun.

Selective Chubs

Those of us in the know have been aware of this for a long time, but just recently the fly-fishing world was thrilled to learn of an immensely important literary project now underway. In a recent issue of *Fly Fisherman* magazine, Nick Lyons announced and outlined in some detail the book he and Craig Woods have been working on for years, *The Masters on the Chub*.

I am fortunate enough to know both of these legendary angling giants. In fact, Craig (who I will call Woody here because that's what those of us in the know call him) and I fished together frequently for five years in Vermont before I moved away. Just the other day he sent me a copy of a bass-fishing article he published in which, bless his heart, he told about me catching the only genuinely bragging-size smallmouth bass of my experience (though I noticed that he didn't feel the need to mention that it was bigger than the one he caught). In the accompanying note he suggested that we needed to go fishing some more because he was running out of anecdotes. As he put it, "What's an outdoor writer without anecdotes? A naked man."

Anyway, as I read Nick's prospectus for *The Masters on the Chub*, and even more as I read Woody's article, I was overcome by an unabashedly self-important sense of history. Years from now, when limited-edition Moroccan-goatskin-bound copies of *The Masters on the Chub* squat authoritatively on bookshelves everywhere, I will be able to look back on my modest but surely braggable part in the creation of that great work. Many was the day astream when I saw Woody pass up Vermont's more trendy sportfish species—suckers, fallfish, garter snakes, snapping turtles—to pursue the wily chub. As I watched him ease stealthily into a good chub pool—water so still it looked about to coagulate—it didn't occur to me that I was a witness to angling history in the making. But then I suppose that's often how it is among those of us in the know; we're so modestly absorbed in our Great Work that we have no time for self-celebration.

But I certainly was involved, and I figure it's not too soon to cash in on those retrospectively heady days and ride the coattails of Woody and Nick to fame as they wrap up their landmark achievement. With a little luck, someday people will speak of us (especially me, I hope) as they now speak in reverent, even ecstatic, tones about the great milestone books of Jennings, Flick, and Schwiebert: "Schullery, yes, let's see; wasn't he the fellow who helped Lyons and Woods with their research?" Or at the very least I can firm up a first-paragraph spot in the acknowledgments section of *The Masters on the Chub*.

And you too, dear reader, can participate in this impending pageant of adoration. Someday, when this epochal book is long in print and the graybeards sit around discussing its genius, you can casually point out that you have a copy of Schullery's article—"You know, the one that started it all?" And if I dare dream large, I can imagine that in some far-off future one of fly fishing's ever-meddlesome revisionist historians might equate me to England's Marryat, who many writers now agree was the behind-the-scenes expert who should have written the dry-fly books that Halford wrote instead: "Well yes, the Lyons/Woods book was a great one, but oh, what a book we lost when Schullery chose not to write on chubs!"

Be that as it may someday be, I gather from Nick's report that it will be a while yet before the book is out. He speaks of difficulties in locating a publisher, but I understand there are other delays as well. I hear (from others among us in the know) that they can't find a fly tier willing or able to produce professional-grade examples of their highly innovative series of "scum flies." There also seems to be some difficulty in getting a sufficiently accurate and heroic cover portrait of a fish whose skin is the color of no known paint. Rumor has it that Dave Whitlock is to be sent to Florence to research arcane medieval paint formulae.

But these delays are to my advantage, as I will have ample time to climb on the bandwagon and get some stuff in print about this soon-to-be classic game species. I will start with the following rumination.

CHUB WATER

Though our British forebears have often discussed the chub, and how and where to fish for it with both bait and flies, you can search many of the renowned American fly-fishing classics—Rhead, Brautigan, Schweibert, Brooks, LaFontaine, Borger, Swisher and Richards, Marinaro, even McClane—without finding more than scant passing mention of chub water. This is a startling omission considering that we all know it on sight, even if our only response in the past was to hold our breath and hastily walk by it in search of something more, well, trouty.

Think how often you've encountered excellent chub water: a turbid and curiously unreflective stretch of stream, just this side of stagnant, that if it moves at all just sort of wiggles. In short, classic chub water, redolent with the promise of the rarified sport only the chub can provide.

And, as Nick and Woody will no doubt soon persuade us, we have too long been harsh and unthinking in our disapproval of the fish caught from such places. Thank Heavens that *The Masters on the Chub* will rectify this injustice and overcome our paucity of appreciation for a great American game fish, whose unique character and many distinctions have been subjected to cruel disdain for so many years.

THE HISTORICAL CHUB

Writing in 1847, George Washington Bethune, one of America's original and most perceptive fly-fishing writers, said of the chub, "He is a bold biter, more ready than welcome at any bait offered him." Soon after, during the American Civil War, the great Thaddeus Norris, one of American angling's most esteemed father figures, summed up this tale of tragic unkindness and sporting neglect:

> *The Chub is a persecuted individual in a Trout-stream; one whose name is cast out as a reproach amongst fly-fishers; whose head is knocked off, or he is thrown ashore on a sunshiny day to linger and die on the pebbly beach, like an Ishmaelite in the sands of the Great Sahara. Every man's hand is against him.*

It is not surprising for a fish that most anglers would prefer not to think of at all that the chub's historical legacy is more involved than most people realize. As it is said that the mosquito was created to make us think better of the housefly, so it might be said that the chub serves a similar role for the trout. How could we have glorified our favorite troutish sport fish for so many years without the "lesser" species like chub to hold up to them for comparison?

And, indeed, there is a darker side to our traditional bias. Perceptive historians have reminded us of the long-forgotten but once-renowned Vermont tackle manufacturer Thomas Chubb, who built a huge factory in Post Mills, Vermont, in the late 1800s, producing countless workmanlike fishing rods for the great many anglers who could not afford the top-end gear but loved fishing nonetheless (the business enterprises of Chubb's competitors Orvis and Leonard could both have fit comfortably in dusty corners of Chubb's building). It is a historical mystery that Chubb's factory burned down and was rebuilt at least twice, which must lead us to wonder if some misinformed local anglers with only the simple rustic's grasp of spelling could have thought that the "Chubb factory" down the road was in fact the source of all those undesirable little fish in the local streams?

Ever since those days, American anglers have taken the same dim view of the chub. Saying "chub" to any gathering of anglers evoked in their minds a sad-eyed little fish—many imagined it as somehow overweight—whose willingness to take a fly they could only wish more trout shared; whose color was the reflection of a Gary, Indiana, sky in an abandoned sewage lagoon; and whose life was a testament to the imperfections of the modern world. The chub, by its unfortunate circumstance of just happening to live in a stream otherwise full of colorfully stripy and polka-dotted trout, long ago fell into that broad category of mercy-exempt creatures known as "trash fish" or—in a powerful colloquialism I once heard from a New England bait fisherman—"shitfish."

CHUBS UNLIMITED

I've managed to catch chubs in well-known trout streams all over the country, often to the complete exclusion of the trout. Happily, this experience further involved me in the Lyons/Woods research, because I have been able to share my own hard-earned insights with Woody as he has developed his pathbreaking salmonid-mimicry theories. After a lifetime of chub-intensive fieldwork, Woody hypothesizes that, setting aside their own chubbish species and subspecies designations, chubs exhibit powerful behavioral traits whose sole evolutionary purpose seems to be to gull anglers into confusing a chub they're casting to, or have just hooked, with a certain species of trout. In Vermont, Woody and I proved to our own satisfaction that in this respect there are thus Rainbow Chubs, Brook Chubs, and Brown Chubs. And my own additional studies in the American West have identified Cutthroat Chubs and, oddly enough, even a few Whitefish Chubs, though that last identification seems to indicate an appallingly low evolutionary ambition, even for a chub.

The place I know them best is Yellowstone National Park, where the chub has revealed the durability and flexibility that characterize the species even while they vex anglers. In 1890, during one of the pioneering fisheries studies in the park, researchers found chubs living with apparent contentment in one of the park's geothermally heated streams at 88°F. Imagine putting a trout in such water; it would boil his overrefined, selective little brain in minutes.

Added to this extraordinary hardiness, chubs are fecund beyond all measures of decency. A twelve-inch-long female Utah Chub carries as many as ninety thousand eggs (a brook trout that size carries about a thousand).

Until now, many of us, denied by cruel tradition the elevated view of the chub that Woody and Nick are about to bestow upon us, would have responded to this startling information with indignation; "Holy Cow! Ninety thousand baby chubs from one mother? No wonder the damn things are so thick everywhere!" To further heighten our indignation, baby chubs innocently eat about the same things that baby trout eat, a direct competition for food that also explains an adult chub's frequent and enthusiastic rises to our dry flies, at least when he can't get scum.

But as I have said, surely this long history of blind prejudice against the chub will end once the world has had the leisure to read the Lyons/Woods book, and American fly fishers will move into an enlightened new era, just as they are now coming to appreciate the finer sporting qualities of the carp.

After all, it would do us no harm to admit that sometimes (come on, I bet it's happened to you, even if you made sure no one was watching at the time), when the trout have become too inflated by their good press to rise to our flies, we've been extravagantly grateful to see a few less fussy chubs rising stupidly here and there.

Some realistic anglers long ago adjusted to the presence and inevitability of the chub, though not to the degree of our European counterparts, who have eagerly fished for them for centuries. Even old Thaddeus Norris, writing so long ago, admitted the chub's sporting worth, at least to the extent of recommending a fly pattern for them. Always the sage, he predicted that we are likely to become fonder of the chub as the years pass because "in many sections of the country it furnishes excellent sport, especially in those streams where Trout have been fished out."

We're not too proud to put a fly over them then, are we?

Besides, the little ones are kind of cute.

Impressionists

ON A DREARY LATE-FEBRUARY MORNING, 1890, WINSLOW HOMER has made one of his infrequent trips down from his home in Prouts Neck, Maine, to New York City. He is in town to visit Gustave Reichard's gallery on Fifth Avenue and see how his showing of Adirondack watercolors is doing and to discuss some other business with Mr. Reichard. When Reichard is called away to the telephone, Homer wanders out into the galleries, which are empty at this time of day except for a slight, tidily dressed, and rather heavily mustached man who, though still in the process of removing his overcoat, has already become intensely absorbed in examining the nearest painting. As Homer watches, the man moves quickly past several paintings, occasionally exclaiming in gusty whispers of obvious delight.

No longer able to restrain himself from greeting such a hearty admirer, Homer crosses the gallery, extends a hand, says, "Good morning to you," and introduces himself.

"The artist himself? Well, this is a happy surprise. My name's Gordon, Theodore Gordon," he said, seeming barely able to tear his gaze away from the art on the gallery walls.

"I don't mean to intrude, Mr. Gordon. Only I couldn't help noticing that you seem to be enjoying the paintings," said Homer.

"Yes! I saw the notice of your exhibition in *Forest and Stream*, and then an angler friend encouraged me to come in. I confess I wasn't sure if he was serious. He was being unaccountably coy and wouldn't tell me why I should come, but I could tell by his insistence that I would be foolish not to follow his advice. And I must say, he was right. I must thank him."

"I will take that as a compliment."

"Of the highest order, sir; your gift dazzles me," he said, practically bursting with enthusiasm. Before Homer could respond, Gordon continued, "Frankly, I have been disappointed, I may say even frustrated, by most artistic portrayals of fly fishing. I will admit that Tait and the others have done pictures that are undeniably very pretty, but they lack some quality, or spirit. Whatever it is, it's just not there. Trout fishing is so much more than they show us. Even the paintings portraying fish on the line, or jumping, are . . . well, how should I put it—your work here makes me realize that the other artists whose angling pictures I've seen manage to be realistic without being real, if you take my meaning. But *these*," he said, indicating with a vague wave of his arm the watercolors around them, "have more than spirit. They have *life*."

"Very kind," Homer said with a slight nod of the head. "It is safe to assume that you are an angler?"

"I sometimes think that is all I am; at least all that matters. Fly fishing especially has become my passion. And tying flies, of course."

Homer's expression brightened. "That being so, I wonder if you have given much thought to . . ." But suddenly distracted, he looked over Gordon's shoulder and said, "Oh, excuse me, duty calls; here are some people to whom I must be excessively cordial. Can you stay a bit?"

"By all means, by all means," Gordon said with an enthusiastic smile, ushering Homer by. "There is plenty to occupy me here."

Homer hurried over to some newly arrived guests, an older and clearly well-to-do couple whom he greeted with almost theatrical warmth. Momentarily free to continue his enjoyment of the art, Gordon moved slowly past two or three more pictures before settling in front of a horizontally elongated scene, *Casting—A Rise*, and studied it with an eager intensity.

Gordon seemed not to notice that Homer was telling his new friends to "Please do come back when you can stay longer," as they shook hands. "I'd love to give you my official tour of the Adirondacks," Homer added, extending an open hand toward the gallery. As the door closed behind them, he returned to Gordon's side, saying, "Now, where were we?"

Ignoring the question, Gordon kept his eyes on the watercolor and said, "You have done some tricky things here, and gotten away with them beautifully."

Homer nodded hopefully, rather like a teacher whose brightest student may have just caught on to something that has left the rest of the class staring dully at the blackboard. "Yes? Go on."

"Well, first, I must say that I had no idea that the whole idea of a trout's rise could be captured so vividly with such a neat little gesture as this," Gordon said, indicating the slight silvery oblong on the dark water. "It's so simple and yet it gives the impression of so much."

"Impression?" Homer said. "Well, yes, that's the word to use these days, isn't it? I might even say that you impress me by using the word."

Gordon seemed nonplussed by this, so Homer added, "Impressionism is the label given these days to a very exciting modern trend in painting, mostly French; it's on my mind right now because these paintings of mine are being called 'impressionistic' in the press. But yes, back to your point, impressions are of the utmost, aren't they? I appreciate your approval, but I don't know that I'd characterize that part of the picture as especially 'tricky.'"

Gordon shook his head. "No, no, sorry; there's just so much in this picture I enjoy that I lost the plot for a minute. It's the line, isn't it? The line's the thing. Your handling of an airborne fly line is so extraordinary that it took me a bit to see past its execution to realize what you've done with it."

Homer, smiling broadly again, said only, "And?"

"It's so graceful, one might even say elegant, unrolling back and forth above the angler like that, that for a moment you don't notice that if it were actually straightened out across the water it would reach at least twice as far away as the rising trout. It's as if you've made the story better by overstating it."

"But that's what we fishermen do best, isn't it?"

Laughing but still not looking away from the picture, Gordon continued. "And the best thing is that it all holds together so well. Even when I realized that realistically the cast is all wrong, I couldn't object. The story's right there!"

"I congratulate you, Mr. Gordon," Homer said, extending his hand to shake Gordon's again. "The past two weeks this room has been several times filled with art enthusiasts—critics, collectors, and all the rest—and not a one of them was angler enough to see that."

"Really?" Gordon said, thought for a moment, and added, "But then, I suppose I'm not too surprised. Fishing is something many of us feel we must claim to be competent at but few have the wherewithal to keep at it consistently or to pursue it deeply, you know?" Gordon looked back at the picture and said, "And the line in the air; how did you ever paint that fine a line? It hardly seems like a brush could come to such a consistently tiny point."

"Oh, it's not paint," Homer said, glancing toward the windows at the gallery's front. "The light's not the best in here today; lean in closer and you'll see. I just scratched it in with a fine stylus. The line's the white of the paper underneath."

"I'll be," Gordon said. "It never occurred to me that artists did that. Painting without paint, you know; I'd have thought that would be against some ancient code of the guild, if you know what I mean."

"Yes, I do." Homer grimaced a bit, apparently remembering something unpleasant. "In some people's minds it may be. But again, as you so aptly observed, art is all about impressions, and the scratched line gives the best impression of the real thing, sliding through the air in that sinuous motion that a good cast has."

"It does that," said Gordon.

"Which reminds me, before I had to go do my manners with those people, I was going to ask you about something. Let's see . . . impressions . . . lines . . . impressions . . . Oh, right, it was flies. I wanted to ask you about flies."

"There at least I'm on my home ground," Gordon said. "Please do."

"Evidently you follow the sporting press? *American Angler, Forest and Stream*, and the rest?"

"Yes," Gordon said, "though lately I'm paying more attention to the *Fishing Gazette*. The English seem to be ahead of us when it comes to thinking about flies."

"Is that so? Well, anyway, I am wondering what you make of the debates—I might better just call them arguments—over fly color. As you can imagine, color is something I give a lot of thought to."

"Ah, the color question," Gordon nodded cheerfully. "Yes, it's one of those happily endless inquiries, isn't it? Gone on forever, that one. Do fish prefer red or purple, or some combination of red and white, and so on? Maybe blue's the thing on Tuesday mornings?" Gordon said. "Frankly, I'm not sure that a debate like that is even asking the right questions."

"How so?" Homer said. "My own experience with trout, at least in the still waters of the Adirondacks, suggests to me that any fly with even a bit of red will usually work better than any fly without."

Gordon seemed to give this some thought, and then said, "All right, let's grant that, for argument's sake. The greater question is, how well does the red fly work, even at that?"

"I don't think I follow you; as I said, it works best."

"Yes," Gordon agreed, nodding. "But is the red fly's best really all that good? I mean, for every fish that takes your red fly, how many casts do you make that no fish takes?"

"Well, quite a few, I'm sure," Homer admitted.

"Just so. I suspect that if you kept count you would find that even with your best fly there are dozens, scores, even hundreds of casts between each fish that takes the fly."

Homer thought about this and said, "Yes, that's most likely true. Some days more than others, certainly, but more often than I'd like. I suppose we tend to forget all those casts because nothing happened. Our memory concentrates on what did happen."

"It does indeed. But do you dare assume that those many uncounted casts didn't work because no fish saw them? Isn't it more likely, in fact mustn't it be a certainty, that a great many more fish that did not want the fly saw it than those few that took it?"

"I suppose so. But that's fishing, isn't it? There are no certainties, no guarantees."

"And be it ever thus, I say! But that's not to say we shouldn't keep trying harder. I can see from your art that you are not only a skilled painter but a gifted observer. You clearly know that little silver ellipse you painted on that picture back there meant that the trout was breaking the surface of the pond to suck down a fly."

"Yes; a mayfly to be precise."

"Right. Good. And doesn't it follow that you'd be best off casting a fly that looks as much as possible like that mayfly? And how many mayflies are all, or even in part, bright red? Not a one."

"Oh, I see; you are, what do they call them, an *imitationist*. I've seen the books; Roosevelt, Scott, and so on."

Gordon shook his head. "Imitationist? I suppose I am. As I said before, I'm not sure what I am. And now that I see your pictures I begin to think that I'm more an impressionist, if you don't mind me applying the painter's language to such a humble art as fly making. But whatever I may be called, I believe in asking these questions. It doesn't seem to me enough to use a fly just because it works a lot of the time. It's more satisfying, and a great deal more fun, to try to figure out *why* it works, and from there to move on to figuring out if something else might work better."

"A noble goal, no doubt, but one thing I've observed about fishermen is that they are fairly hidebound; I fear you will have a lonely uphill battle persuading them to abandon their Red Ibis and their Parmacheene Belle for something less grand and flashy, no matter how well it imitates a living insect."

"The Belle, yes, it's a good example," Gordon said.

"Of what?"

"Of everything that's problematic with our thinking about flies. Look here, at this." Gordon led the way over to another small watercolor entitled *Leaping Trout*, which showed two airborne trout, both seen broadside; the left one already descending toward the water and the right one nicely curved at the apex of its jump. Near the upper edge of the picture, left of center, a single ghostly mayfly can be discerned, though probably not noticed by the casual observer.

"Here," he said, "you can clearly see the brook trout's fins. Now, on the one hand, the Parmacheene Belle is said to be imitative. According to Wells, who apparently created it, we are to believe that it imitates the belly fin of a brook trout. Not that it does, of course; as most people make the pattern, it's even lacking the black inner band."

Homer agreed, adding, "Yes, I've heard about that, and it has always sounded a bit absurd to me. It begs the bigger question of what a brook trout's belly fin is doing wandering around in the water without the rest of the trout attached to it."

"Right," Gordon said, warming to his subject. "That's the other hand."

"Yes, but hold on," Homer said. "This may be about color after all, and I believe we have wandered back onto my ground. Couldn't it be argued that of all the physical features of a brook trout that might attract a larger trout to eat it, perhaps those strikingly colorful fins are, what shall I call them, well, the *triggers* that move the larger fish to action? And if that's so, might we wonder if the visual trigger is compelling enough that they'll go for it even without the trout? What does a fish know about the behavior of the creatures it eats?"

"Quite a lot, I suspect, but I grant your point," Gordon said.

"But that's only part of, what did you call it, 'the color question'? And I wonder, are you familiar with Chevreul's work on color?"

"No. Chevreul? Sounds French; a fly tier?" Gordon said.

Homer shook his head. "Not a fly tier at all, though he worked with matters of vital interest to fly tiers everywhere. And painters are profoundly indebted to him. I surely have been." Homer paused, as if trying to decide where to start. "Actually we should begin with Buffon . . . , but no, let that go. Chevreul is enough for now. Michel-Eugène Chevreul was a brilliant French chemist who worked in industry. He worked with dyes of all sorts and eventually wrote several books on the use and identification of colors, primarily for industrial purposes, like the production of reliably colored fabrics and many other materials. Stained glass, for one thing. Come to think of it, he worked a lot with yarns, which would be of most immediate interest to fly tiers, of course."

"When was this?" Gordon said.

"He died recently, just in the past few years, but his books on color were published, oh, forty or fifty years ago. You should look into his work; I think for your purposes it would be invaluable. Let's see, his titles were a bit windy; if I were you, I'd start with *The Principles of Harmony and Contrast of Colours and Their Application to the Arts*. There's an English translation around. It has a strong bearing on your color question. Now that I think of it, the Parmacheene Belle is ripe for discussion in light of Chevreul."

"Please, go on. Though I do hope I'm not taking too much of your time," Gordon said, glancing toward the gallery office.

"Mr. Reichard probably won't miss me for some time yet; his new telephone threatens to become his best friend," Homer said with a smile. "In fact, why don't we sit down," he added, leading Gordon to a pair of chairs from which they could still see several of the paintings.

"Now," said Homer, settling back, "Chevreul. Being constantly involved in the world of colors in the most intensely industrial and yet scientific way, the man noticed things that, if they had not escaped the attention of the occasional perceptive artist before him, at least had not been articulated in any public manner, much less published with such intellectual discipline as he was capable of. One of the most important results of his work was what he called the law of the simultaneous contrasts of colors, and I can think of no more perfect illustration for demonstrating that law than the brook trout's fin, or for that matter the wing of the Parmacheene Belle. While working with commercial dyers, Chevreul and his associates were frequently disappointed with the accuracy of the colors requested from the dye makers for this or that product. No matter how precisely and consistently the makers prepared the dyes, the resulting colors didn't appear right in the final product, whether it was woven fabric or a carpet. But it was Chevreul's genius to realize that it wasn't the dyes that were at fault. It was the human eye. Think of the Parmacheene Belle, with those brightly colored married wings, reds and whites joined immediately against one another."

"Yes, that's usually what we see first of the fish, even if it's in deep water," Gordon said.

"And with good reason. Chevreul showed that the eye not only discerns that strong difference, the contrast between the two colors if you will; it involuntarily heightens it."

"What, like an optical illusion?"

"I suppose you could call it that, though that oversimplifies what is happening. I am trying to summarize in a few words what has seemed to me at times an almost infinitely complex matter. Chevreul was brilliant, but it appears that even he may only have opened a conversation that still results in progressively more abstruse theorizing about color by other experts." Homer paused, puzzling over how to go on. "Let me try to generalize here. At its simplest, imagine two pieces of shirt pasteboard. One is black and the other is a medium gray. Chevreul noticed, and demonstrated beyond any doubt, that seen individually they will give you a different impression of their shades than if you place them side by side. Fundamentally, all the variations in colors and shades, and all the theorizing about those matters, could be said to radiate out from that simple reality. And they permeate our viewpoints of our world; fashions in clothing, in wallpaper, in furniture, in everything including trout flies. Think about fashions; there's always talk about complementary and contrasting colors; if a man's tie 'works' with his suit and that sort of thing, you know. Sometimes, as when two garish colors are combined in a woman's coat, we say the colors 'clash,' but most of the time our visual and mental reaction to contiguous hues or colors operates on much more subtle, even, it might be said, subconscious levels."

"But back to the Belle, if you don't mind," Gordon said. "Even if our perceptions of that fly's wing are affected, or intensified, by all this—and I am fully prepared to believe they are—how can we know that the trout's eye is likewise fooled? We do know their eye is biologically not all that unlike ours, but just as certainly it is not identical."

"True enough, and in discussing the red and white of the fly pattern we've neglected that narrow streak of black that separates them; does that heighten our impression of the contrast between the two colors?"

"Could be. Makes sense," Gordon said. "After all, nature put it there for some reason."

The two men sat in thought for a moment, and then Homer said, "I'm tempted to say that, no matter how you or I perceive the Parmacheene Belle, the success of the fly pattern in catching fish is all the proof we should need that it works similarly for the trout's eye. But if I said that then I suppose you would point out that on any given day almost any pattern might happen to be successful, so I will just admit that I don't know just what's going on."

"Nor do I," Gordon said. "Still I believe that we must soldier on despite the imponderables, and this Chevreul sounds as if he was on the right track in ways that may, indeed, help. Even if all his studies do is make us pay more attention to the combinations and intensities of the colors we choose for our flies, that would be quite enough to be grateful for."

"I don't doubt it, Mr. Gordon. We are, after all, not dealing with precise visible qualities; we are dealing with impressions. If you choose to continue your quixotic search for some higher form of accuracy in trout flies, I imagine it would be well to keep that in mind."

"I have already decided to do so," said Gordon, abruptly rising from his chair and hurrying back to *Leaping Trout*. As Homer followed him, Gordon said, "But there's more to this picture that interests me."

"Oh yes?" said Homer.

"You've obviously seen a great many trout jump, so I'm telling you nothing you don't know here, but when any fish jumps it's a wildly messy process. Look here," Gordon pointed to the fish in the picture. "There would be water in the air in all directions; these trout would still be shedding water from their sides. More than that, a jumping trout, whether fighting a hook and line, or chasing some insect, or just jumping for whatever reasons of their own, rarely achieves this perfection of form. They twist and tumble ass-over-teakettle, if you'll pardon the expression. You've left that out of all these jumping trout pictures," Gordon said, pointing to some of the other similar paintings on exhibit.

"Exactly correct," said Homer, but sensing that Gordon had more to say, let it go at that.

"Now, I do understand that as a professional artist you're about a good bit more than capturing all that wildness and energy. You're also celebrating the beauty of the fish in a way that a variety of people, including potential buyers, can admire."

"Yes," Homer agreed. "There are compromises to be made even in art. Plenty of them, in fact."

"But, and I hope I'm not overreaching propriety by asking this, but isn't it a bit of a . . . I'm trying to think of a diplomatic term here . . ."

Homer smiled and said, "Perhaps 'betrayal' is the word you're looking for?"

"Yes, sorry, but that's the word." Gordon agreed. "Isn't it at least a bit of a betrayal of nature to leave all that excitement out of the picture? You undoubtedly have the wherewithal to paint a vivid impression of what a trout's jump is really like, and I'm sure you could do it without compromising your portrayal of the trout's beauty."

"I could," said Homer, studying the picture as if imagining the very brushstrokes that would lay in the trailing stream of water extending back and down from each fish's tail and suggest the sparkling drops of random spray nearby. "I suppose I'm still a slave to fashion when it comes to the airborne fish. These perfect profiles of fish are what people expect. Kilbourne may have established a standard for showing a trout jumping; no splashes or other interference with the details of the fish's appearance." He gave the idea some more thought, then said, "Now that I think of it, it could also be that in the back of our minds we artists are all still stuck with the responsibilities of the first naturalists' portraits of trout. Certainly our fish are in quick motion rather than mere dead corpses laid out flat like, say, Doughty's, but maybe we're still trying to fulfill the obligations of a specimen illustrator."

"Well, you know your business, but as a casual observer I wouldn't take it that far," said Gordon. "What you've done with these trout makes Kilbourne's fish seem like cartoons. Water or not, these are real fish over real water. They lack nothing, in detail or in, well, the impression of life."

"Thank you. That degree of success is my fondest hope, of course."

"As it is mine with trout flies, though when I see these paintings my hope seems by far the more modest ambition."

"Oh no," Homer objected, "not at all. It's all the same search, isn't it?" He paused again, looking like a professor preparing a lecture. "Look at it this way. The mistake people make when they talk about impressions, or impressionism, as confused as that term has become these days, is in assuming that an impressionist painting is merely a vague version—and a much simpler version—of what they like to call 'the real thing.' But that's perfectly wrong. In its way, and I mean in the way that we *see* it rather than in the way that a fine camera might capture it, the *impression* is the real thing. The impression turns out to be what is most precise; it's the whole of what you see, not the specifics of what is there." He took Gordon by the arm and led him to a somewhat larger painting, *Sunrise, Fishing in the Adirondacks*. "Look here, at this one. The Taits of the world would try to give you a thousand details of that hillside, painting in all the trees, the branches, even the individual twigs and leaves. But in the dim light and motion and excitement of a moment at dawn, who sees all that? *This*," he waggled his hand at the picture, whose hillside background was a long mass of dark, suggestive shapes and shadows, "*this* is what you see. At least I'm sure it's what *I* see."

"Yes, me too." Gordon nodded in agreement.

"And isn't that what you're after when you tie a fly?"

"I hope it always will be. Like you I have to think of the market sometimes; most fly fishers still love their Belles and their Ibises, but that's no excuse for me to settle for them myself. And I don't mind admitting that if someday I can do in my art what you've done in yours I will be a fulfilled man."

"Very kind," Homer said, with a slight bow. Looking toward the office, he added, "And now, I suspect we both may have other business to attend to."

"Indeed I do, but I have one more question about the trout in this picture," he said, indicating *Leaping Trout*. "What's the story with this mayfly?" He pointed to the faint silhouette of the flying insect at the top edge of the painting. "Are we to assume that these trout have attempted to reach it, but couldn't jump high enough?"

"That's as good a story as any, isn't it?" Homer said.

"Well, yes, but . . ." Gordon paused. "Is that a part of impressionism too? That its stories aren't necessarily clear?"

"Not really. At least not any more than any other kind of painting. Artists have always been tempted to put little mysteries into their pictures."

"For me," said Gordon, "the real mystery isn't whether or not the trout were trying to reach the fly; after all, the fly is well positioned relative to the fishes' trajectories for that to be a perfectly plausible explanation of what's going on. No, the bigger mystery is the way you've painted the fly. It looks transparent. It looks like a ghost."

"Hmm," said Homer, "you're right. Actually, I didn't want it to be too well defined. It would have competed overmuch with the trout for your attention. But for you, being a fisherman, it did anyway. Most people don't even notice it."

"Really? Well, then, have you ever considered painting mayflies for their own sake. I mean, as the subject of a picture?"

"Not really. There's a couple butterflies over there," he said, indicating another picture of a trout appearing to jump toward, but just below, a pair of large butterflies, "but, sad to say, I doubt that people would be interested in a picture of a mayfly."

Gordon shook his head in disagreement, then said, "If *you* painted it they would." With that, he shook Homer's hand, turned, and headed for the door. Homer watched him pass through the door, then lingered in the galleries for a while, looking at the paintings and finding himself seeing them through the eyes of that remarkable angler who had just left.

Eventually, Mr. Reichard emerged from his office and joined Homer. All business, Reichard asked him if the man he'd been talking to was a likely customer.

"I think not," said Homer, realizing that throughout their conversation it had not once occurred to him to wonder if Gordon might buy something. "Interesting fellow though."

Reichard, accustomed to making expert readings of the people who came into his galleries, said, "Ah, an artist then?"

Homer gave this some thought, then said, "Yes. Yes, that would be more like it. And if I'm any judge of these things, he is a great one in the making."

Teepees over Texas

I. The Stage

This first part is true; well, as true as any fishing story ever is.

For a couple years in the mid-1950s my family lived in Corpus Christi, Texas, well down the Texas coast. We were fresh transplants from Pennsylvania, from where my dad, a Lutheran minister, and my mother, a sometime substitute teacher and full-time housekeeper/child-raiser/loving mom, had uprooted us when Dad felt the call to go into what was then called "domestic mission work," that is, moving to some corner of the country that the higher-ups in the Lutheran Church saw as ripe for the establishment of a new congregation. It was arduous work and a dramatic change from his distinguished pastorate of a prosperous, well-established parish in Hershey. In seeming contradiction to the challenges and difficulty of the work, the assignment came with a disgracefully large cut in salary. But it was where his faith and his heart led him, so we went.

For my brother, sister, and me, Texas was a radically new and nearly exotic environment, a "faraway place" in the best sense of the song: in our mid-1950s Corpus Christi suburb there was a house a few doors down with an actual producing banana tree in the backyard; everybody talked funny; everybody thought *we* talked funny; we were "Yankees" at a time when most southerners were still pretty sure the Civil War wasn't over yet; there were "horned toads" in the field across the street; summer was very hot; and there was no discernible winter. Perhaps most revealing of what an unknown country Texas was to Pennsylvanians at the time, when my second grade teacher back in Hershey organized my little classmates to send me letters, one of them asked me if I rode a horse to school.

The fishing was also a big change from Pennsylvania. My dad, though not an especially enthusiastic angler himself, was a genuinely great dad—the founding hero of my life. But my big brother Steve, then an adolescent of twelve or so and already my other hero, was a passionate fisherman, so Dad took Steve and me fishing regularly. I was about seven, and I think now that I probably got to go along because I would have whined insufferably if they'd left me behind. I may have lacked Steve's informed enthusiasm for fishing, but I managed to have some fun in my own way.

I can now see that I had no idea what a wonderful, broadening experience life in Texas was for three kids from the North. Ever since then, as I've lived in various states and as I studied and wrote about our country's history and natural history, that brief period along the south Texas coast has come productively to mind countless times. This is especially true of the fishing, which I now remember with considerable wonder, though at the time it was just "fishin'"—most definitely without the final *g*.

These little outings typically began with a brief stop at a local fish market where Dad would buy a frozen squarish half-brick-size block of near-jumbo shrimp chunks, our bait for the day. This I recall with something approaching horror—that we would squander what today would be a high-value culinary target by chucking it piece by piece into the Gulf of Mexico in hopes of catching fish that wouldn't taste anywhere near as good as the bait.

We fished along the Gulf Coast shore somewhere not far from town, out toward the Naval Air Station that was, at that time, home of the famous Blue Angels. I suppose that one of Dad's parishioners or perhaps some neighborhood informant had put him on to this fishing spot. The shore consisted of rock and sand and some straggling vegetation that I don't remember any better than that, and the land was restrained by a low wall (wood? concrete? I couldn't say) so the water was a couple feet or so down from where we stood to cast.

And there were lots of crabs. I was both fascinated and a little creeped out by them as they scuttled around with their fine disregard for any sense of going in the direction they were facing.

We all had the standard working-class rods and reels of the day. I suppose that because we were just *fishin'* rather than *fishing* we probably called our rods poles, but as I write about them now, I can't seem to break the habit of calling rods *rods*. Dad's and mine were solid fiberglass and Steve's was, as best he can remember, one of those steel ones that were popular at the time. Steve also reminds me that Dad's rod was white, and mine, which Steve (bless his heart) saved all these years and just recently mailed to me, was a sort of sickly green.

The sturdiness of this gear was demonstrated to me when the rod arrived at my door. At some point on its trip in a mailing tube from Michigan to Montana the rod managed to poke through Steve's careful packaging, exposing a couple inches of the tip to the vicissitudes and violence of shipping, but the tip was just fine. I suppose that the fates, which surely would have made short work of any modern name-brand rod in such a situation, either found it was too tough to break or figured that it was too cheap to bother with.

Dad had an Akron Pflueger bait-casting reel, and Steve and I had lesser-quality (read "cheap, but what could be afforded") reels. At the time I saw these reels as godawfully vicious things, which I guess they were for a kid like me, because of more or less constant backlashes. I did not come to appreciate the better class of such reels as the beautiful little machines they were and are until many years later. In a separate package Steve also sent me Dad's reel, so I can authoritatively state that our lines were multicolored Dacron ones, probably about 30-pound test; at every stage in our gear we were evidently going for strength rather than subtlety. We even had a landing net, long-handled enough to reach down to the water from where we stood and retrieve whatever it was we'd hooked.

Almost any fish we caught was exciting. Casting from shore like that our harvest was not huge, but it might include some version of saltwater catfish, small black drumfish, saltwater "sea trout," eels, and something that looked like whiting. I suppose there were other species, but the very few surviving photos of Steve and me holding up a stringer of fish are blurry and hard to interpret. I don't remember ever catching anything that was more than a foot long unless it was an eel, which I refused to handle, but we happily filled the stringer with the more select species and brought them home. My mother, a rural Ohio girl highly skilled in the breading and frying of northern bluegills, perch, and catfish, expertly prepared them for us. This was *fishin'*!

As I say, I was seven, which meant that beneath my cheerful, Sunday-School-perfect-attendance, bowtie-wearing veneer, I was just another soulless little barbarian. It is true that I lacked some of the finer instincts for cruelty of some of my new seven-year-old pals, who would have instantly grasped the opportunities for sadistic anatomical experiments on those crabs, but my blithe obliviousness to good sense did show itself in other ways, most involving whatever fish was unlucky enough to take my bait.

Being a bright and inventive boy (i.e., inherently dangerous), and having immediately but not quite accurately figured out that the whole point of this enterprise was the bringing-in of fish, I took a value-added approach to any fish I caught that might best be called "catch-and-catch-again." Having successfully hooked and summarily dragged in some poor benighted little thing, I saw no reason to stop there, so I'd haul back and wing the still-hooked fish as far out as I could and reel it in again. And again. And maybe again. I was perceptive enough to notice that the excitement level dropped off rapidly with each successive retrieve, and I do have a faint almost-memory that my brother or my dad eventually noticed this bit of childhood savagery, but I really couldn't say how they reacted. Probably not positively.

(Here's a thought, though. The great angler-conservationist Lee Wulff, one of the pioneers of modern catch-and-release fishing, taught us that a sport fish is too valuable to be caught just once. Given that, I suppose that to a nonangling alien who had just stepped from his spacecraft my catch-and-catch-again behavior might seem qualitatively indistinguishable from modern catch-and-release enthusiasts who—though leaving out the stupid part where you cast the same fish out again—do, as a group at least, happily catch the same fish again and again.)

But my most important memories of our fishing there on the edge of the Gulf of Mexico involve my dad. I'm sure he would have been my hero whatever he might have looked like, but he made it even easier by being physically heroic, a big strong man from a family of big strong men. He grew up in the mountains of eastern Pennsylvania, where my grandpa was an engineer on the coal trains. In the mid-1930s, after graduating from high school, Dad and a friend converted a huge old Studebaker into a home-made coal truck (being young, they painted "The Bomber" in tall uneven letters on the sides) and spent a year as bootleg coal miners, illegally and dangerously working the leftover veins in deep shafts that were abandoned but still owned by the big companies. For the rest of his life, any of his parishioners who happened to look closely at his hands during the communion service might have noticed the coal scars. Luckily, a helpful and generous uncle, recognizing that this big kid had the makings of something special, funded his first year of college, where he thrived; his extracurricular activities included playing left tackle on the football team.

I mention Dad's size and strength because it comes up in any discussion of his fishing. It was partly because of his inherent ath-leticism that he naturally decided that he'd like to cast his bait a lot farther than his tackle would normally allow. After all, on several occasions at various places along the coast we'd seen the topmost bits of very large fish, apparently sharks or tarpon, much farther out from shore than we could reach. Dad would have just reflexively asked himself, "Why not those, too?"

As it happened, among Dad's amazing array of constructive pastimes was the making of lead soldiers, probably an interest he'd picked up while young and stuck with. I remember him pouring the bright molten lead into the molds in his little workroom in the basement. He made and then carefully painted hundreds of the little soldiers, cowboys, and Indians that we kids played with. In the thirty-some years since his death, this little lead army has been entirely demobilized, but I still have one piece, a flat and roughly triangular representation of an Indian teepee two inches high, two inches across the base, and a tad less than a quarter inch thick. In shallow relief on the front an apparent Indian of unknown tribe is seated in the folded-back doorway. On the reverse in equally shallow relief are the stitching lines of the various hides used to make the teepee.

These little teepees were Dad's spontaneous answer to the problem of casting farther. He drilled a hole near the top, tied one on his line near the hooked shrimp, and let fly. The little Akron wasn't made for such heavy work and must have protested vigorously at the untoward exit velocity of all that weight, but being well made it did keep working. I'm sure that there was much added excitement to be had just in reaching that much farther into the ocean.

It is with the thought of that excitement that I move from the what happened to my cherished dreams of what might have happened. Or, better said, what *should* have happened.

In other books I have explored some fairly ethereal concepts of the just rewards an angler's worthiness might reap. To summarize, I cherish the thought that there could be some greater justice, some karmic scorekeeping, at work when we fish; that once in a great while some fisherman, through hard work, good intentions, and whatever other worthy qualities might affect such things, is given a fish by a benevolent Nature for reasons beyond anything to do with said angler's fish-savvy, location, or the properly managed presentation of bait, lure, or fly; that there might be a higher rightness of that angler catching a fish right then. Laugh, oh ye skeptics, but I believe—and I'm sure many others share this secret conviction— that I may have seen it happen. And in any event, it is a great thing to hope for a day like that. Besides, even if it didn't happen I can still wish that it could.

II. The Story

Such was a hot bright day there on the very western shore of the Gulf of Mexico. And I wonder if what happened that day may not have been merely about the worthiness of my dad, as profoundly worthy as he was. It could instead have been the result of the cumulative and slowly gathering worthiness of countless dads on countless little fishing trips where they exercised inexhaustible patience and good humor to put their fussy, overexcited, and beloved children in the way of catching a fish. Up to that day, dating from some distant past when it had happened last, on such trips and shores beyond counting, much that was inexpressible and incalculable was earned and totted up in some ethereal but long-abiding Great Book of Merit—earned by all those hours of rain, sunburn, snagged lines, tearful setbacks, hooked fingertips . . . all those gallons of big and little anglers' finger-blood, worm juice, and other forms of fishing goo reflexively wiped on pant legs and shirtsleeves beyond counting. Some mighty and vigilant presence, however each of us may choose to define him/her/it in our own spiritual dictionary, kept imperturbable watch over the filling of that book, looking to the time for the eventual paying-out of a cosmic, balance-restoring dividend. And that day along our anonymous bit of Texas shoreline, the time had come.

There were two or three other fishing parties strung every fifty or so yards along the shore in both directions, and we could hear their occasional hoots and hollers as they enjoyed their beers or brought in a fish. We were having at least as much fun, though minus the beer, and we'd been at it long enough to have a fair stringer-load of this and that. Steve, who almost always caught the most fish, landed three of the fish we knew as trout. I'd caught one or two luckless, anonymous little fish, messed around with pretty rocks and shells to my own personal satisfaction, and only once had to edge discreetly away from the random, mysterious wanderings of the crabs. Very little of our time had been wasted picking at backlashes, which I always did with the pointless but inextinguishable hope that I'd find the one loop that when pulled would magically release the rest of the snarl.

So it had already been a good day when Dad muscled out a particularly lofty and hard-driven cast in the general direction of Tampa. Steve and I often paused to watch in both awe and envy as his shrimp/teepee combination arced far and high over the water, landing with a respectable splash, so we were still watching as the splash was followed, within a second or two and at the very same spot, by a rolling bulge of the water.

This startled us, but Dad, always prompt to begin his retrieve, was already cranking the reel when the fish hit. Feeling the take, he set the hook in an unyielding weight. The white rod bent, then bent some more without any evidence of shoreward progress on the part of whatever he'd hooked way out there. His surprise immediately turned to a greater interest than he normally showed while fishing. He smiled.

The fish did not share his enjoyment of the moment. Line instantly poured from the reel so that Dad, though trying to thumb down the whirring reel, had to ease off or risk a seriously abraded thumb. He told us later that he had no idea how much line the reel might hold, much less if the fish might take it all, but before he could deliberate on this previously unimaginable eventuality the fish jumped.

This was perhaps the moment of greatest revelation for us all, the visual peak of the thrills still to come. Out of the sea came this enormous furious shining thing, seeming to polaris vertically into the air, twisting and spraying water whose particles flashed as if the fish was shedding small glittering fragments of its own glowing substance. Though we were too shocked to yet realize it, and were hardly qualified to identify or estimate such a thing, Dad had just hooked a tarpon of about sixty pounds.

Being who he was both professionally and personally, Dad did not let out with the obligatory string of frantic and jubilant expletives that would be expected from almost any mortal angler at such a moment, so you'll just have to trust me that his slowly spoken "Gee whiz!" carried into the salty atmosphere all the rhetorical heft and authority of any of the less savory alternative outbursts with which others might have honored the moment. He'd never hooked a big fish in his life. He smiled again.

I was frozen in a mixture of fear and wonder; that big thing out there scared me well beyond thinking. But Steve, seeing the determination in his father's face and sensing what Dad was about to do, shouted, "Just keep the line tight, Daddy! You're gonna have to *play* him! You can't just reel him in." Which was of course precisely what Dad—who knew no better and had instantly accepted this violent invitation to find out who was stronger, him or this grand fish—was about to try to do.

I have no doubt that this advice saved the day. Steve, having even at that age accumulated priceless experience with big fish while wrestling porcine carp ashore from a big Ohio reservoir near our grandma's house, was well ahead of Dad in his fish-landing savvy.

I give Dad a lot of credit for instantly recognizing the superior knowledge in this matter of the small, earnest boy standing next to him. He said, "Okay, so I just hold on until it gets tired?" Come what may, this had now become a family effort.

"Reel in when it lets you," Steve said, with what I thought was amazing calm, "but keep ready to let it run."

That was what it came down to, all right, but as anyone who has hooked a powerful fish knows, it never feels that simple when it's happening to you. The fish was off again as soon as it landed, but this time it ran across in front of us rather than away from us, thus keeping about the same distance from shore. This provided Dad with a moment to get used to keeping the line taut, giving a little line while keeping the rod tip well up to make the fish work, and at least starting to think about reeling in when next it seemed possible. A second equally breathtaking jump turned the fish around, and, as it headed back to the general neighborhood of its first jump, I unfroze enough to squeak, "What *is* it, Daddy?"

"It's a tarpon, and a goddamn big one!" This came from one of the nearest fishermen, a tall, slender man with short but bright red hair who, having noticed the commotion out in the water, ran right over and was standing a few yards off to Dad's right. "What'd you get 'im on?"

"A chunk of shrimp, Frank," Dad said, without the least loss of concentration on his line cutting through the water.

The man had apparently been too focused on the fish to take a good look at Dad, but now he said, "Oh, sorry Rev'rend, I didn't see it was you in your civvies." Frank was an irregularly attending member of Dad's new flock. He might even have been the one who put us onto this spot in the first place. He said, "Have you caught tarpon before?" just as the fish launched itself in another noisy, torquing leap, then headed for deeper water, peeling line from the reel in uneven jerks as Dad gave the reel spool as much pressure as his thumb could stand.

"Not that I can recall right this minute," he said, still smiling and not taking his eyes off the line.

Frank smiled too and said, "Well, if you don't mind some advice, next time he's not runnin' it would be a good idea if you set the hook good and hard a couple more times. There's not much in their mouths for a hook to get hold of, and you want to make sure he's on good." He looked around at Steve and me, took in our little tackle box and suddenly inadequate net, then turned back to Dad and added, "You're gonna have to work him right up against the boards here so's we can grab him." Steve and I looked at each other, both hoping that by "we" he meant himself rather than either of us.

The big events in our lives, at least our memories of them, come down the years to us along a complicated path of gradually refined and embroidered memory and an equally problematic conversational process in which the story is told and retold until it finally settles into a more or less "official" form that passes for the truth as long as no participant in the original event feels too slighted by its final and often minutely tailored version.

By this admittedly imperfect route, over the years our family has come to agree that the tarpon jumped eight times and made at least a dozen progressively shorter runs over the next twenty minutes; that the skin on Dad's thumb was worn through and bleeding by the time he was sure enough of the fish to simply crank it across the last twenty feet of open water and nearly to our feet; that finally Frank, fully but very casually dressed, unhesitatingly eased himself over the edge into the waist-deep water, saying, "All right, Reverend, all you gotta do is lead him past in front of me"; that Dad hauled back one final time on the straining rod and did just that; that Frank crouched down under the fish, cradled it in both arms, lifted it, walked it to shore, and dumped it on the dry ground in front of us; that it would be forever uncertain whether it was the baited hook or the teepee, which had somehow gotten irretrievably wedged deep in the tarpon's jaw or throat, that gave Dad the necessary hold to play the fish; or, for that matter, if it was the bait or the teepee that the fish wanted to eat in the first place; that Frank's wife Loretta brought her camera down and photographed Dad, Steve, and me holding the fish across in front of us (well, all of the holding was done by Dad, but Steve and I put our arms under the ends and did try to lift, just for the form of the thing); that Frank, having assured us that tarpon "tasted like crap warmed-over," suggested we just roll it back into the water rather than haul it somewhere to be weighed and photographed some more; that after the big splash settled and we watched with some relief as the tarpon righted itself and swam slowly off, Frank said, "I reckon I might should be comin' to your church more regular"; and that the next day we found Loretta's picture of us with our giant smiles and Dad's well-earned fish on the front page of the local paper under the headline, "Local Pastor's Miracle of the Fishes."

III. THE POINT

I have no idea how long it takes for the Great Book of Merit to refill. Moreover, I go back and forth on the equally important question of whether there is only one such book that all of us are cumulatively trying to fill, or if there are instead many such books—even if we might each have our own. The idea of such books, or one cumulative book, goes a long way toward explaining the cherished notion of fisherman's luck: for instance, the bewildering way that once in a great while some blessed young angler catches a monster on his first cast. It couldn't be the number of casts he made that earned a fish that fast, so maybe it was the long accretion of casts by anglers everywhere that happened to reach some critical luckiness mass just as the kid lobbed his lure into the water.

Luck is by most accounts an enormously complicated idea. One of my favorite of Ernie Schwiebert's fishing stories, one he told in *Remembrances of Rivers Past* (1972), involved his youthful fishing trip to a Norwegian river famous for its giant Atlantic salmon, where he caught what any angler would consider the fish of a lifetime, a fifty-one-pounder. When an older friend, a lifelong salmon fisherman, heard of the catch, he put the achievement in perspective for Ernie:

Killing a salmon like that at your age is tragic, he smiled philosophically. *It's like stumbling over the Grail on your first mission from Camelot.*

Such is luck and, perhaps more to the point, such is fishing. Judging from the embarrassingly and entirely undeserved kindnesses that Nature has bestowed upon me in the past seventy years, I remain certain that even if at a desk in some implausible ethereal realm an accounting of each fisherman's worthiness is indeed being kept, it seems undeniable to me that someone besides me must be doing something sensationally meritorious to make up for my own inconsequential fumblings as an angler. Nothing else would explain all the magical things that keep happening to me whenever I go fishing.

Still, as grateful as I am for whatever mysteries were involved in giving me all those rich and astonishing days astream, I would trade them all if Dad could just have really caught that tarpon so long ago.

Holocene Outfitters, Inc.

EDITOR'S NOTE: THE EDITORIAL BOARD OF *TIME AND AGAIN: THE Journal of the Amateur Temporalist* is extraordinarily pleased to present this interview with Ernest Waters, chairman of the board of Holocene Outfitters, Inc., the only time-travel firm currently authorized by the International Time Travel Regulatory Commission (ITTRC) to lead nongovernmental expeditions to the past. Holocene Outfitters is well known for its fishing trips to various times past, but until now their confidentiality policy has kept them from divulging any details of what have become known as "dream trips for the angler who has everything, especially an obscene amount of money." We think Mr. Waters's pithy and forthright responses to our questions will interest our readers intensely, considering the extremely closemouthed position of federal agencies and international time-travel managers in recent years.

TAA: Let's start with your services. What do you provide your clients?

EW: Well, as our ads say, they get "the fishing trip of a million lifetimes." They get to fish where no sportfisherman has gone before.

TAA: And where would that be?

EW: Oh, pretty much any where, any time, over the past twelve thousand or so years.

TAA: It being that open, how do they decide?

EW: Ah, I see—it's a human nature question: what does the man who can have anything actually want?

TAA: More or less.

EW: I suppose there are two types of fishermen among our clientele; though, considering the cost—and you must know there are plenty of small nations whose treasuries could not buy their president a trip like this—I doubt that these people are typical of your average fishermen.

The first type is pretty introspective about it all. I call them the Thoreaus. These are the people who simply want to immerse themselves in some truly unique environment and fish there. They love it when they catch something, but they're far more interested in soaking up the moment than in getting their picture taken with some incredible fish, which is what seems to drive the second type, whom I call the Hemingways. These are the guys whose passion is right on the surface, guys who only want to go back where, or I should say when, they can catch a fish bigger than anything available on earth today.

TAA: They don't even care what kind?

EW: Uh, yeah, you bet they do. Very few of them are interested in a fish species that doesn't exist anymore; I suppose because it doesn't give them anything to compare with today. They want a bigger bass, or salmon, or whatever, than any of their pals can catch today.

TAA: Which group is harder to please?

EW: They're both easy. Where we take them, we can't lose.

TAA: Do you mind telling me where that is? Or when, I mean.

EW: No, I don't. As it turns out, it's changed over the years. When we first advertised, a lot of people didn't understand about the blackouts.

TAA: Why did that matter to fishermen? I would have thought they just wanted to go sometime so long ago nobody would be around.

EW: No, not at all. A lot of the fly fishermen, for example, wanted to fish waters they'd read about in the old books—you know, like fishing the Dove in Walton's time, or the Catskills when Gordon was around, or, even more recently, the Rivière Arnaud with Campbell. But all of that was out of the question, of course.

TAA: I suspect many of our readers don't know much about the blackouts; would you mind reviewing that?

EW: Sure, as much as I know, anyway. About fifty years ago, or whenever the feds actually figured out time travel, they had a lot of trouble with casualties at first. People would come back looking like they'd spent a couple hours in a microwave, almost unrecognizable except that their clothes kind of held them in the shape of a person. This wasn't a secret at the time; you could check the newsviews if you're really interested. Almost right away, though, they also noticed that these problems happened most often with the more recent trips, like someone trying to go back just to 1800 or so. Hardly ever did anyone who went back more than five hundred years have trouble.

TAA: But I thought it had to do with going back to where your ancestors were.

EW: That's the other correlation they picked up, of course. You're a lot safer sending someone of German descent back five hundred years if you send them to Asia rather than to Austria. It seems that the closer you get to your own genes, the greater the risk, even that long ago. But three hundred years ago, even four hundred for most people, there's no place on the planet where you're not likely to get a little scrambled, no matter how far away your ancestors were at the time. Quite a few of those early travelers came back with no apparent physical damage, but a lot of their dominoes didn't have spots.

TAA: I think it was the physicist Norton Cronon who made that famous wisecrack about how this gave a whole new meaning to the term "relativity."

EW: Right. Whatever it is that causes the problem, the more recent the relatives, the greater the risk. That's probably because as the human population has grown, each of our extended families has grown, too. The more ancestors there are around, the more likely you are to have a problem. It's just the odds. So the blackout was imposed on time travel to all dates more recent than about 1500, and supposedly it's been in effect ever since.

TAA: Supposedly?

EW: Well, you have to understand that we're a minor concessioner in an industry dominated by giant governments, but we do hear the occasional rumor. Some say there've been technological advances that will allow more recent stops. Other rumors make you wonder if there isn't the occasional suicide run, just to get something important straightened out.

TAA: What could be that important?

EW: Just about anything. Back when this time travel thing all started, there were a lot of people who wanted to go back and just clean up the whole mess we'd made of history; you know, some surgical manipulations, like assassinating Hitler as a child, or giving John Wilkes Booth's arm a little bump, that sort of thing. You know politicians; can you believe the powermongers would ever give up on wanting to fine-tune the past to their own advantage? But on the other hand, the establishment is built mostly on the bad things that have happened, so there would naturally be a lot of resistance among the winners to changing the past. Like, the last thing the modern Christian Church would want is for some goofy vigilante SWAT team to materialize in Judea and rescue Jesus from crucifixion; I mean, what would you do with him *then*?

TAA: I know we're getting kind of far from the subject of fishing, but doesn't it worry you that someone might be mucking around with your past?

EW: Why should it? If they manage to do it, I'll never know. I'll just have grown up and lived my whole life in a slightly different world. I almost said, "a slightly different world than what it might have been before," but of course there won't have been any "before." I'd rather they didn't screw around like that, but I don't expect I'll have a chance to stop them.

But you were asking about the fishermen, and what they wanted.

TAA: Oh, right. We were talking about their preferences, and the blackouts.

EW: If it were possible, some of these people would have paid anything to spend a day fishing with Charles Cotton, say, in the 1600s. We couldn't do that anyway, because our primary rule is not to even be seen by the locals, much less talk to them, so we never knowingly take anyone close to people of the past. But this limitation was a real disappointment to these guys. Some of them wanted to fish the Madison River in Montana before it was cleaned out by that series of bio-catastrophes in the late 2020s. Only trouble was, they didn't realize that what they really wanted was to fish it in about 1920—to catch those early generations of browns and rainbows that had only been introduced in the late 1800s and grew like crazy. So they'd ask to go back to fish the Madison a thousand years or so, but when they got there they'd discover that the river was full of native fish, the cutthroat trout, grayling, and whitefish, and none of them was nearly big enough or braggy enough for someone who'd just paid a fortune to catch them. We got some bad press out of that, but it did teach us to give these guys a history lesson as part of the application process.

TAA: So you do a lot more trips to Europe now?

EW: Well, after a couple years and some lawsuits that were very expensive, even though we were eventually exonerated, we figured out what they wanted, and actually, no, we still mostly do trips to the United States. Word has gotten around about the dangers of going back to a time and place too close to your ancestors, so most of our clients tend to shy away from the Old World even though we assure them that there's virtually no risk in fishing Europe five thousand years ago. There are a few people who want to catch the big huchen in the Rhine or the giant salmonids of Lake Baikal, but they're nothing compared to the people who want to catch the Pyramid Lake trout—they're real monsters. The trout, I mean. On the other hand, there are always people who choose the north shore of Lake Superior, or Maine, or Labrador. A thousand years ago, the Rangeley Lakes area of Maine had enormous brook trout. For a trout fisherman, a picture of himself with a twenty-five-pound brook trout seems as good as with a sixty-pound Pyramid Lake Cutthroat.

TAA: What do you tell the ones who want to go back further than you're allowed?

EW: No.

TAA: I assumed that. Please don't think I was prying, or suggesting you might break the law. I just wondered if a lot of people are interested in ancient fishes.

EW: Sure they are. We have a waiting list for the earlier epochs, in case the rules are ever relaxed. It would be raw magic to lead an expedition back eighty million years or so, to catch what the paleoecologists are calling the "Dawn Trout." They just identified it a couple years ago and it's not all that big, but imagine the social cachet of being able to say you caught the *first* trout. But I doubt that the blackout will ever be lifted. Too many butterflies, you know.

TAA: Butterflies?

EW: You *are* young, aren't you? The blackout on the earlier periods, say prior to the Pleistocene glaciations, was imposed even before the one on recent centuries.

TAA: I knew that.

EW: Did you know why?

TAA: I just assumed it was because it was too dangerous back there. You know, dinosaurs and dire wolves, and all that.

EW: Yes, I suppose a lot of people assume that. And in a way, it was. The deal was, it took a long time before they figured out *how* dangerous, and in the meantime, we may have lost a lot of things we don't even know we should miss. Or, on the other hand, maybe we gained. Who knows now?

TAA: Some readers might consider those statements incoherent.

EW: Well, you kind of made me jump ahead of myself. Once you've been involved in time travel a little while, grammar seems a lot harder than you thought.

What happened was this. Though they recognized that it was vitally important to keep the past and the present from influencing each other, they didn't really know how to stop it. At first, all they really worried about was catching some awful disease. Paleomicrobiology has come a long way, but really, who knows for sure what all sorts of nasty little organisms infected the animal life millions of years ago and what effects it might have if we inhaled one and brought it back with us?

TAA: But there were all those disinfectant procedures, and isolation periods.

EW: Right, and they worked pretty well, but we still didn't know how bad it could get, at least until the Varnisson Expedition, let's see, that was about 2059. Varnisson led a team of paleontologists back to the late Paleozoic to do a little genealogical observation on the earliest dinosaurs.

TAA: Those were the guys who sneezed?

EW: That's the popular lore, yes, but that's mostly just a folkloric muddle involving some of H. G. Wells's stories. It *is* true that one of them was just getting over a cold, and may have sneezed a couple times, but we don't really know what happened. They were all wearing the full-body prophylactic suits and masks, after all. Anyway, when they got back to the present, everything seemed fine until a few hours later they started noticing things, and when they finally got together and sorted it out, they discovered that an entire family of small dinosaurs and quite a few mammals and plants they remembered were no longer in the textbooks. Over the next few years, they made a list—it finally included more than twelve hundred species—of things that didn't seem the same as they remembered them. There was apparently a whole group of birds missing; they called them finches.

TAA: Couldn't they go back and fix it?

EW: How? Retract the sneeze? You can't predict your landing time very precisely in those gigantic time-jumps, and this was one of the biggest—250 million years, give or take a few millennia—so they couldn't just plot a course to the week before they'd arrived on their previous trip, leave a note to themselves, and hope it didn't get eaten by something before their *earlier* expedition showed up a week *later*.

TAA: I'm beginning to see what you mean about grammar.

EW: Of course, on account of the blackout, they couldn't go back to their own office the week before they left and warn themselves not to go. You may never have heard about this, but in those really short time jumps, like a few weeks back from the present, you can't even transport something inanimate, like a note that will just come to rest on a desk, without frying the head of whoever is operating the machine. Apparently there's no amount of shielding that protects against whatever that effect is. When that path opens, it reaches beyond the transport chamber.

Anyway, the Varnisson team taught us how easily we could mess things up, from a great distance. Or from a great time, I mean.

TAA: It's okay; I know what you meant.

EW: Anyway, those Varnisson guys had to rediscover their whole world, and it didn't go well. There were half a dozen of them, and not one of them stayed in science. Two of them became evangelists. It wasn't that the world was different overall; it was still a very complex place. What with evolution being so busy, all the niches were still full of animals of some sort or another, but remember that these guys weren't familiar enough with most plant or animal orders to even know what might have changed. Two of the guys, including one of the evangelists, finally killed themselves. One left a note saying he just couldn't get used to lavender cats.

TAA: So what was different about the butterflies?

EW: The butterflies? Oh, I see. No, you misunderstand; the butterflies weren't part of the problem with the Varnisson team. I was using them to make a point you didn't get. Back in the 1960s, when science was first fooling around with computers to try to predict events, a meteorologist named Lorenz cooked up this little fantasy to suggest how complex climate is. He used the example of a butterfly whose wings were stirring the air in Beijing, saying that a month or so later, those tiny air currents the butterfly launched would result in a storm system in New York. It was called the butterfly effect. It was just a poetic metaphor for a very complex process, but it suggested to their little minds what a complicated place the world is. They didn't have a clue, really, but they were groping toward some understanding.

Anyway, after the Varnisson disaster, it became clear that the butterfly effect was as much temporal as spatial; the further back you mess with things, the greater the changes in your present. Even if what you did was really, really tiny, if you gave it enough time it would magnify into spectacular things. Well, the Varnisson mess proved that millions of years were more than enough time, so there were only a few deep-time expeditions after it, and then all that really old stuff was made off limits.

TAA: As far as you know, you mean.

EW: Yes, there's always that uncertainty. Powerful people don't change, and we don't really know for sure what all might have gone on secretly. But remember that it was about this same time that Cronon launched consequence theory, to try to track change a little more meaningfully. It's always been largely a theoretical science because right after he published, the blackout was imposed, so he couldn't test his ideas by actually traveling back in time any more, at least not far back enough to make a difference. Shortly before his death, in fact I think it was when he accepted the Nobel Prize for his development of the first eventuality loop model, Cronon said he wished he could go back a billion and a half years and just spit in one of those primordial algal hot pools where life first got stewed into existence, then zoom forward to the present to see if the modern world would look anything like he remembered. Someone pointed out to him that if he did that, he'd probably change the future so much that he'd vanish on the spot.

TAA: Your mention of pools reminds me that this is supposed to be an interview about fishing. Why don't you describe one of the most popular trips?

EW: That would be Lake Bonneville—Ancient Lake Bonneville, maybe twelve thousand years ago, before it shrank down and left the little puddle we call the Great Salt Lake. Next time you're in Salt Lake City, look at those steep hills to the east; hundreds of feet up the slopes you can still see the old shorelines. There wasn't a lot of paleontology to suggest that there were huge fish there, but we heard about them from some geologists who had made a government trip and claimed that they saw alligators swimming around; turned out they were just really big trout.

We take parties of two fishermen each back there with a small modified bonefishing skiff, and fish for cutthrout trout that average forty pounds. Our record is ninety-three. I was the guide on that trip.

TAA: Ninety-three pounds! That's like a tarpon.

EW: Yes, it was. That fish had a whole swan in its stomach.

TAA: So maybe a seventy-five-pound trout, normally.

EW: After it digested the swan, I guess so. The record weight of ninety-three still counts though.

TAA: I'm sure that the anglers among our readers would love to hear more details. What kind of lures do you use?

EW: No lures. This is a fly-fishing trip. Almost all of these people use big tarpon flies, ten or twelve inches long, and 14-weight rods and lines. The fish are really strong, stronger than any modern saltwater fish that size. We have to use custom-made reels with at least eight hundred yards of backing because these fish make longer runs than any modern fish I've ever seen, fresh or salt.

TAA: That's more than half a mile of line.

EW: Right, and sometimes they clean you, just take it all, and when they hit the end of the line and you see the bare reel spool, there's not even a hesitation; the line just pops and is gone. I don't know why they're so strong, and every biologist I ask doubts it's even possible until I show the pictures.

TAA: Ancient Lake Bonneville was a huge inland sea. How do you find fish in water that big?

EW: Well obviously, because we're using fly tackle, we can't be trolling three hundred feet deep or anything like that. We have to find the fish near the surface, which is where they spend a surprising amount of time. If we're lucky, we can work the shoreline, especially off the mouths of the big streams, in the spring when the fish are spawning. That's a sight to see, let me tell you.

But often we use our sonar to find shoals out in the middle of the lake. We've been there often enough now to have some good maps, but we don't usually go back to just the same time, and the shoreline and the bottom keep changing, so we have to keep learning.

TAA: Why do you say "if we're lucky" you fish the shoreline? Why don't you just fish wherever you want?

EW: Because there are all these other people around. Anybody who believes there weren't many people in North America ten or twelve thousand years ago hasn't spent any time cruising through that time trying to find a quiet place to fish. It's a damn nuisance, but we have to keep our effects on those people to an absolute minimum. A close encounter with a bonefishing skiff with three or four guys in it could have some pretty weird consequences in aboriginal culture, you know?

TAA: Like what?

EW: Well, nobody's sure of course. It just seems like the sort of thing we had better avoid, even being seen from a distance. That's how the ITTRC feels, anyway, so we work hard at it.

TAA: How do you avoid encounters, then?

EW: Fishing is from dusk until dawn. We all wear full-spectrum booster lenses, the temporary implants that dissolve after a couple days, so it looks like daytime to the sports, and in the dark with our neutral-reflection camo we're hard to see from much of a distance. The fish are looking up at a lighter sky, so they can see to feed just fine. I think they're mostly nocturnal predators, which might suggest something about how much fishing pressure the local people were putting on them, even back then.

Sometimes we do get lucky, though, and can fish in broad daylight. Now and then we can even go ashore and fish a stream. An Italian client once caught a forty-two-pound cutthroat on a dry fly in a little creek; he spent his entire ten-hour visit landing that one fish and was thrilled. Sport is a very complicated thing, different in the eye of each beholder.

TAA: Has a party ever been in danger from the local residents?

EW: We had a couple dicey moments before we worked out the specifics. When you're sent back, you're all sitting there in the boat, you know, and the pilot only has about a minute to respond if he sees that he's going to come out in the middle of someone's beach party.

TAA: You mean that you can see where you're going to come out before you are actually there?

EW: In essence, yes. Or, no. You're there, but you're not, uh, materialized yet. There's some kind of delay in the process—someone called it Heisenberg's Hiccup—right after you've reached the designated time, when you're kind of hovering right over the water but haven't become visible yet. It's been a really lucky break that it happens that way, because the pilot can just shift a few hours or days and find a nice quiet spot.

We usually try to come out in the middle of a big body of water, far from shore; less chance of running into the neighbors.

Anyway, once one of our groups emerged about twenty-five yards from shore, and just as they did a whole bunch of people came strolling out of the bushes carrying canoes of some sort. There isn't any noise when you arrive in the past; just suddenly you're there. It isn't like there's a big boom that would scare people. And once you're there, it takes a few minutes to power up to jump back to the present, even in the most desperate emergency. Once your initial jump is complete and you're out of the hiccup, there's only enough power for the return jump, so it's either stay in that spot or go home. So this one time, there stood a couple dozen Paleoindian hunter-gatherers face-to-face with the pilot, the guide, and a pair of Swiss financiers. Nobody said a word. The locals' eyebrows kind of went up a little, one or two of them smiled, and the pilot just hunkered down as low as he could, signaled the others to duck out of sight, and with as little engine noise as possible he backed the boat straight away until they were a mile or so from shore.

TAA: Did they feel threatened?

EW: No. He saw a couple of things that might have been atlatls, but he didn't see any threatening behavior. I sometimes think that for those people, thousands of years ago, seeing that boat must have been a lot like a UFO sighting would be for people today; it's really interesting, but other than its conversational value, what can you do about it? It's so unusual it's almost meaningless.

TAA: It could work its way into the local belief system somehow.

EW: That's true, and I'd feel bad about that, but with so many little isolated tribes and groups around, the odds are apparently pretty small that whatever belief it inspired would last very long, or reach very far. It doesn't seem to have done so, anyway. I mean, nobody's ever found a petroglyph of a bone-fishing skiff or anything.

Don't get me wrong here. I don't want to seem uncaring. We go to extraordinary lengths to avoid those encounters. Besides, the procedures we put clients through to keep them from carrying any diseases back in time are incredible. But the big reason that these trips are allowed to the recent past is that even if we do mess something up, it's too close to the present for evolution to be screwed up before time reaches us today.

Oh, it's possible that we could have some effects. If we got in a fight and killed some guy ten thousand years ago it could mess up some major genealogy among Native Americans later, but remember that in the first couple hundred years after 1492, the descendants of these Paleoindians were all but wiped out in North America by European crowd diseases—smallpox and all that—and then the few Indians that were left had their cultures and their genetics scrambled some more by the various so-called Indian wars that were really just purges, and by the reservation system. That was all a horrible crime and tragedy, but it did sort of cut off the rest of us from getting caught in any of the prehistoric effects today's time travelers might have had on them.

TAA: Uh, right. I guess. But back to fishing. You must have a lot of people who want to fish for big game, like marlin.

EW: We can't help them. I don't care what you've heard otherwise; there really is a size limit on what can be transported back in time, and we can't get a decent-sized boat for deep-sea fishing through. But we don't hear much from those people anyway. I think their sense of competition almost requires them to fish in the same modern waters, for the same fish. Their idea of a trophy is different, I guess, and going into the past must seem a little unsporting to them. We do take the skiff to saltwater, though. The shallow-water fly fishermen don't seem to mind competing in the past.

TAA: What for?

EW: Well, at first there was a lot of interest in bonefish. They're fairly small, ten or twelve pounds is a really good one, but that's all a matter of scale, like with brook trout. People figured that there must have been some time in the past when all the world's bonefish populations weren't being hammered by so many fishermen, that there were bigger ones, like twenty-five-pounders. A twenty-five-pound bonefish isn't really impossible today, but if it exists it's so rare that it might as well not. I think some of our customers figured that a twenty-five-pounder would be kind of like the Holy Grail, a fight so breathtaking that they'd achieve some higher sporting state of being.

But after a few trips, we discovered that bonefish were always about the size they are now. The average was a little larger, of course, probably because the habitat was in better shape, but there weren't any giants. So then we started getting more tarpon fishermen.

TAA: But tarpon is a big-game fish.

EW: It sure is, but it's also a shallow-water fish. Fly fishermen have been catching them for a couple hundred years now, since the 1890s at least. The tarpon has the charming and accommodating habit of spending lots of time in shallow water where these guys can pitch a big streamer in front of it on a light leader, and actually land it. It's a religious experience for them, and they're always trying to get a bigger one.

TAA: How big?

EW: There have always been a few guys trying to catch the first 300-pounder on a fly. They've increased their record ounce by ounce, so that they're up to 277 now, but it's kind of like the four-minute mile was for so many years. It just seems impossible.

Anyway, it didn't take these guys long to figure out that a thousand years ago, before all the destructive overfishing and trashing of the ocean, tarpon might have been significantly larger; their ecology isn't much like the bonefish's. So then we started getting lots of tarpon trips.

TAA: But would the modern record-keepers credit a prehistoric trophy?

EW: Certainly not. The International Game Fish Association (IGFA) has been quite firm on that point. But even if they were willing to allow a prehistoric fish into the records, we couldn't let it happen. We don't even allow people to *kill* the fish, much less bring them back. Everything gets released, right after it's measured and photographed. We use self-cauterizing barbless hooks, and we rarely see a fish swim away that we think isn't going to recover from the trauma of the fight.

TAA: But if they can't get credit for the trophy, why do they bother?

EW: I told you, sport is complicated. These aren't the Pop-eye-armed deep-sea fishermen; these are fly fishermen. They pride themselves on their lotus-eating tendencies, and even the most competitive ones, the Hemingways, have their gentler moments. They still get *personal* credit for doing it, and they still get to show the holovideos to their pals at whatever club they belong to. And I understand that the IGFA is considering opening a new Hall of Prehistoric Fame, just to showcase pictures of ancient fish being held up by maniacally grinning fly fishermen.

TAA: Have they caught a three-hundred-pounder yet?

EW: No, but they've lost several that I think would have gone three hundred, for sure. Fly-fishing tackle has come a long way in the past one hundred years, but a three-hundred-pound fish is still right along the threshold of what the equipment can be made to handle. It still comes down to that light section of tippet attached to the fly.

Frankly, I think that the first three-hundred-pounder is more likely to be caught in the present. What with the climate having warmed so much, the tarpon's effective range has expanded a lot, and some of those new habitats look very promising.

TAA: Care to predict?

EW: I think that the Chesapeake Sea, and especially the Maryland Archipelago, has created the perfect place to grow three-hundred-pound tarpon. All those submerged agricultural lands are going to make terrific baitfish habitat. Pretty soon, too.

TAA: Well, Mr. Waters, this has been great fun, and I'm sure our readers will be as grateful as I am to finally get the lowdown on Holocene Outfitters.

EW: Happy to oblige.

TAA: I do have one last question. Something you said earlier just struck me. You don't allow your clients to kill any fish, right?

EW: Correct. The ITTRC conventions absolutely forbid transportation of pretime life-forms, even dead ones, so there's no point in killing a fish. The decontamination of people and gear before and after the trip would be an exercise in absurdity if the people brought a whole prehistoric animal back with them.

It was only because we could practice catch-and-release that we weren't dismissed out of hand when we originally submitted our proposal to the ITTRC for a time-traveling sport-fishing concession. The trophy hunters were laughed out of the room because their sport depends so much on bringing trophies home. They wanted full-mount saber-tooth lions and mastodon heads in their dens.

TAA: But didn't you say there was a swan in the stomach of the biggest trout you caught?

EW: Right. Big one, too.

TAA: Well, how did you know it was a swan in there if you didn't kill the fish?

EW: It was simple enough. The trout's stomach was distended like there was something big in it, so the pilot got hold of the fish and held it down in the boat for a better look. It was a huge fish and still quite strong, so he had to use all four hands to keep it still, but once he got it quiet I just reached my forehand into its mouth all the way up to my second wrist and grabbed it. The trout had swallowed the swan butt first, so I got a good grip on the head and just pulled the whole thing right out. It was dead, of course. Then we put the fish back in the lake.

TAA: What did you do with the swan?

EW: Oh, we just took some pictures and pitched it back in the lake. And you know what? That big trout came right back up and sucked it down again. Whoever would have thought it would be interested in eating, so soon after going through all that? Oh well, that's cutthroats, you know?

Me and Joe Went Fishin'

I
OPENIN' DAY

ME AND JOE WENT FISHIN' FOR TROUT ON THE OPENIN' DAY LAST weekend. It weren't too bad, and there was even a few minutes when the snow stopped and another time when I could feel my feet.

Every year for about a month before openin' day, we piss and moan about doin' this. The water's always up to our goozles and so muddy it has chunks, and the fish are all hunkered down under the ice and roots and rocks with their eyes closed. Worst of all any more, the easy places to reach are all shoulder to shoulder with fancy-dan fly fishermen—Joe calls 'em the "Pastel Hatch"—so we have to traipse across some sloppy pasture to get to a stretch of water with too many trees for those boys to show off their backcasts.

Which ain't all bad, now I come to think about it, because most of the real trout are back in there too, not generally wishin' to socialize with the tank trout that the hatchery fellas dumped in at all the bridges the day before. But as fishin' weather goes, it's not what you would call your halcyon Waltonian angling idyll.

But we know we have to go, what with openin' day bein' one of those, what do you call'em, perennial rituals—like grubbin' out the rain gutters in the spring before the first big rains, or goin' to church on Christmas Eve. Some things you just do.

Still, last Saturday was way too lousy outside for us to want to dig our own nightcrawlers from the manure pile, and since Joe's leftover worms from last year had all froze up harder than a whore's heart in his garage back in January when Doreen accidently bumped down the thermostat on the slow cooker, we went down to the hardware and got a couple dozen from Herman. I took along some corn, and though I could tell that this rankled Joe a tad when he saw the Green Giant can on the dashboard, he didn't say nothin' at the time.

As expected, there was a pack of fly fishermen in full regalia fifty yards either way from the road bridge. We didn't want to give anything away by walkin' around them where they might see us goin' upstream to the good spots, so I drove on. This worked out pretty good, because a quarter mile past the bridge we could see that Oscar Ferguson was keepin' his bull in the barn on account of the weather. So I pulled the old Power Wagon over into the weeds without blockin' his gate and we cut across his pasture to that big S-bend in the river that's shaded over so heavy all the way across that even when you're wadin' you sometimes have to duck your head just to get through to the next hole.

As we was dodgin' frozen meadow muffins in the pasture and the snow was showin' signs of turnin' to rain, I said, "Y'know, for as much cash as those fly guys must have to afford all that stuff they fish with, how come they can't buy a good razor? They all look like they shaved three days ago."

"Yeah," said Joe, "but have you noticed? They look like that every day."

"Y'know, you're right. How do you suppose they manage that?"

"Turns out I know this one," Joe said, with a bit of pride in his voice at being informed on this particular cultural issue. "They do it a-purpose!"

"You're shittin' me!" I responded in surprise. "You mean they *want* to look like bums?"

"Nope, no shittin' involved whatsoever. Havin' a three-day growth is the goin' fashion in their world. Has been for quite a while, apparently. It's what you might call a fundamental shift in the groomin' paradigm for young men of a certain socioeconomic strata to look like that."

By then we'd reached the river, and before I had time to work up any more indignation about the ratty-looking boys, Joe pointed with his pole across the water and said, "There's usually a good one hangin' fairly low in the water just back of where the roots of that black locust kinda flop over into the water. See that bit of a curve in the current in there, just where the ice stops? But you got to get it down to him quick or he'll let it go by."

"That ain't no black locust," I said.

"It ain't?"

"Hell no. Look at them old leaves hangin' on there," I said, wavin' my pole to a couple of the lowest branches that hadn't let go of their leaves over the winter. "Black locust's leaves is pinnately compound and anybody with the sense God gave a goose can see that that tree's got leaves that's palmately compound." Besides," I added, kicking up some snowy mud with my wader boot, "this here's the wrong kind of dirt. Black locusts do real good on your

Inceptisols and Ultisols, and this here ground's got too much clay. Nope; that there's a horse chestnut."

"Huh," he said, squintin' a bit at the tree. "S'pose you're right, now I take better look. Well, there's a trout in there anyway. Why don't you go first?"

It was a tricky cast. As Joe had advised, I had to lob my night-crawler gob up high enough to go a foot or so above the roots so it'd land in the little backwater and swing down and around, clear into the main current without either hangin' up on the shore or catchin' on the roots. I usually like a few easy casts to start the day, and tryin' to make a bullseye like this fresh out of the bullpen always gives me the jitters, but I managed it okay and a trout grabbed onto the hook and made the mistake of goin' downstream rather than into the roots, so I was able to get him in pretty quick.

As I was whacking his head on the butt of my pole I said, "So how do they do it?"

"Do what?" Joe said.

"Those boys back there. How do they always look like they shaved three days ago?"

"Oh, right. Well, Doreen, you know, she reads up on this kind of thing down to the beauty parlor, and she says now you can buy shavers where you can adjust the blades so they leave all that hair on your face."

"What, you mean like raisin' the blade on your lawnmower?"

"Same difference, yup," Joe said, movin' a few feet downstream and keepin' his pole tip low so's it wouldn't hang up in the branches right above us.

"But why?" I said, slidin' my trout into my back coat pocket. "Why would anyone who's goin' to the bother of shavin' in the first place not want to finish the job? You gonna try behind that rock?"

"Beats me. Probably comes under my Unified Theory of Off-spring Behavior, which requires that every new generation has to do as many things as possible that will annoy their parents." Then, answerin' my other question, he added, "Yeah, but if I remember right, this time of year, what with the water bein' so deep and all, you really need to get the gob to land just over on the far side of the rock, without the line getting hung up on the rock but just washin' over it so that the current on the near side will grab it and drag the gob right around the downstream side of the rock," and havin' said it, he did it.

"Good call, I'd say" I allowed, as a trout considerably larger than mine took hold and bolted downstream but didn't stand much chance against Joe's 20-pound test and was soon on its way to Joe's canvas sack.

As he was whacking its head on a big rock by his feet, he said, "Oh, and they even got a name for it."

"For what?"

"The scraggly beard. They call it 'designer stubble.'"

"Silliest thing I ever heard of."

"No doubt, but you know, you got to give them fashion types credit. It's one damn good piece of salesmanship, after a few thousand years of a fella either havin' a beard or not, to find somethin' new in between to sell steep to the style minded."

Before I could even consider this commercial angle on the whole thing, Joe said, "Hey, here comes one now." He pointed his pole upstream, and sure enough, through the branches a ways upstream I could see a bristly young fella in the standard fly-fishin' getup movin' real slow downstream. He was in the water, just workin' his way around the rocks and roots along the other bank, and bein' so low in the water he hardly had to duck under the branches at all. Sometimes he was near up to his chest as he eased in and out of the deeper spots, and he was holdin' his fly pole backwards so it stuck out straight behind him low over the water. He hardly made a stir in the water, either, and I could tell Joe was takin' this in.

"Knows what he's doin'," he said, quietly.

"But how's he ever gonna get a fly in the air with all them branches just above his head?" I was nearly whisperin' now, not wanting to mess with the fella's chances.

Just then he froze where he was and didn't move for at least two minutes, and so we didn't either, though my coffee was puttin' in for a transfer and it wasn't the kind of day when standin' still was all that rewardin'. Then he reached real slow down to his reel and fiddled with it for a second. I could see that he'd run his line out the length of his rod and hooked the fly into the reel frame somewhere, and now he was lettin' it loose, because suddenly the line dropped into the water at his side. Then, without movin' anything but the hand that was holdin' the rod, he made a little flip and the rod reached out low across the water and so did the line, and I could just see what must have been a big fly make a little plop up against a rock along the near shore. Nothin' happened for a second, and then he hauled back on the rod, all the time keepin' it just a foot or so above the water. There was a big commotion in the water just down from the rock, and we could tell that he'd hooked a trout bigger than either of ours. He horsed it over to him, reached around behind him where he had a net hangin' down his back, got it under the trout and lifted, and we both let out some air as we saw that this fish was a *lot* bigger than either of ours.

We both were just sure what was gonna happen next, what with these fellas bein' such high-minded purists and all, but we was wrong. Instead of recitin' a few stanzas of Baudelaire to the fish and lettin' it go, he reached into his vest and brought out a slick little blackjack sort of thing, I kid you not, and gave it a good one on the head without even takin' it out of the net. Then he put the cosh away and, holdin' the net under his arm with the fish still danglin' in it, he reeled in, turned toward the far bank, climbed out of the water, and disappeared through the trees.

For a few seconds we just stood there gawkin' upstream and shiverin', until Joe said, "Well I'll be dipped."

II
RESERVOIR CRAPPIES

Me and Joe went fishin' last Saturday. Got a call from the brother-in-law that the crappies was bitin' up to the reservoir, so I calls Joe and tells him we ought to go. After Joe doin' some negotiatin' with the little woman that I could overhear pretty clear even though he thought he had his hand over the phone, he come back on and says he'd hook his johnboat up and pick me up first thing in the mornin'.

So we got to the ramp just about sunup and, once the boat was uncranked into the water and he cussed the kicker into action, we went straight over to that deep channel just off the big bed of lily pads there along the east side. We figured that from there we could always try for some bass if the brother-in-law turned out to be full of bullpucky, as he is about half the time any more.

But the crappie fishin' turned out to be okay. In a half hour I picked up five or six keepers and Joe maybe two or three more than that, until things slowed down and we had a spell to catch up on current events.

"Hope you don't mind my sayin, but I couldn't help overhearin' Doreen on the phone last night," I said. "Seemed a tad overwrought."

"Yeah, well, she's been pretty agitated about the daughter, off at that college and all," Joe said, untwistin' the thermos cap and pourin' himself some coffee. "You?" he said, pointin' the thermos my way.

"Sure," I said, fishin' a fairly dry Dixie cup from under the seat and shaking it out good so's he could pour me some. "What's goin' on? She havin' boy trouble?"

"Nah. Wish it was that simple. It's the same thing as your boy had that one year, remember? When he got to navel gazin'?"

"Oh that. Existential emergency, huh? Sorry to hear."

"Yeah, the whole schmear, you know? Who am I, what am I here for, what is truth, what can be known, what's the mind, where's the seat of the soul, on and on and on."

"Who's on first," I said, shakin' my head in sympathy.

"Right," he said, "That ol' Cartesian dualism gets'em every time. You never heard such agonizin'. She calls home all worked up, and she's got Doreen goin' to the books just to try to keep up. Kierkegaard and Bultmann stacked right there next to the Yellow Pages." He thought about it for a minute, then added, with justifiable outrage in his voice, "And of course if there *is* a boy sniffin' around—and it ain't at all unlikely, you know, considering her, well . . ."

"Her 'thatness,' you might say."

"Yup," Joe said, shakin' his head with a sad widening of his eyes, "she sure has grown into her thatness these last couple years, hasn't she? Anyway, when a girl like that gets herself stirred up with all this panicky intellectualizin', it just plays into a young fella's agenda."

"No stoppin' that," I said.

"No, guess not," he agreed. "That's boys and girls anyway, ain't it? Just so long as they keep their paws off *my* daughter," he added, ironically. He made another cast, wrung his free hand in exasperation, and said, "It's just that these kids, I mean, they don't need these gratuitous aggravations. They show up at college damn near clueless about everything and next thing they know some Distinguished Professor of Deep Left Field gets ahold of 'em and just opens the metaphysical floodgates on 'em."

"That Shaw fella, he had it right about teachers," I said.

"I often thought so," Joe said and started a retrieve that seemed to me a bit too rushed, but before I could mention this he went on. "*You* know how all this goes," he said, takin' his eyes off his bobber to look my way. "You was through the same thing with the boy."

"Yeah, but Hell's bells," I said, wincin' a bit at the reminder, "Hosmer wasn't no freshman when it hit him. He'd had two or three years of, well, truth be told, tomcattin' around chasin' girls and such, to kinda, you know, find hisself a little. And even then he just fiddled around with some rationalist epistemology and a little transtheism now and then, kind of on the weekends. Never did quite go off the deep end."

"No?!" said Joe with surprise. "What about that time he took off for Tibet? What do you call that? Where the hell he'd end up?"

I winced again. "Oh, well, yeah; true enough. Sorry, I guess I try to blank that whole episode out of my mind. Yeah, it was Singapore. Hoz never was that sharp on the geography. Had to wire him airfare to get back. The missus about had kittens. Fool kid hadn't even packed spare socks. But you're right. He couldn't handle it neither, even then."

I reeled in to check my nightcrawler. It was gettin' a little pale the way they do, but it was still all there so I chucked it back out about five yards farther than before.

Joe picked up a couple little ones quick together, and as he was tossin' the second one back and I was lightin' up a smoke, I said, "But Joe, I still think this thing with your girl sounds worse than it was for Hosmer. Pardon me for sayin', but Krystal's mighty ripe in *many* regards at her age, not just her whatchacall, wholesome amplitude. If she ain't careful she could get herself mired so deep in the whole ontology swamp that she never sees daylight again. End up a professor herself, like as not. That's no life for a nice girl like her."

"Yeah, I know," Joe said, reelin' in his line. Again it seemed to me he ought to have left it set a little longer, but I figured I'd best not say anything right then, and he said, "I guess I'd almost as soon some boy got through to her. At least that has a certain biotic inevitability to it." He finished reelin' in his line and took a look at his hook, which was bare now, then stared off toward the far end of the lake and said, in a resigned sort of tone, "Give me the good ol' evolutionary imperatives over normative ethics any day." He flipped his mostly unsmoked cigarette over toward the lily pads, and I must've looked surprised about that, so he said, "I promised Doreen I'd only take a couple puffs off each one. Hand me the can, wouldya?"

I passed him the Folgers can and he dug a big crawler out, thumbnailed off an inch or so, dropped the rest back into the can, threaded the bit he'd kept onto his hook, and lobbed a cast just off to the left of the deepest part of the channel.

"Good idea," I said. "We haven't tried on that side."

We fished a while longer with no action, and then I said, "But y'know, Joe, even if she does manage to stay on track here at first, once she's on that particular reflective railroad most likely her next stop's gonna be nihilism."

"Yeah," he said bitterly, finishin' off his coffee and screwin' the cup back on the thermos with considerable energy. "That bastard Nietzsche! I'd like to kick his ass up on his shoulders. Didn't he have *any* kids, for Hell's sake?"

"Not so's anybody noticed, apparently. Maybe it was just as well. Can you imagine bein' a kid raised by a flickerin' bulb like that?"

Joe laughed. "Nope. Mrs. Nietzsche'd have to be some kind of postmodern ninja hero saint to handle him."

The sun had got pretty high by now. The crappie water and the lily pad bed was both out in the sun, so I suggested that we might want to move down the lake to the cattail flags up under the big maples and see about some bluegills, but Joe said Doreen wanted him to paint the trellises alongside the garden, so we decided to call it quits.

As we was crossin' back to the marina Joe yelled to me over the racket of the kicker, "We might should try again earlier tomorrow, don't you think?"

"Sounds good to me," I said. "I been thinkin' of gettin' some of them fluorescent chartreuse spider jigs that Herman's just got in at the hardware. I got a hunch there's some bigger crappies back there than we been seein', and maybe a change of menu would appeal."

Joe hollered back, "Pick me up some minnas while you're at it, wouldya? The big shiners, if he's got any."

"No problemo," I said.

III

SELECTIVE SNAKES

Me and Joe went fishin' over the long weekend. I reckoned that with three days and some change we had time enough to make it worthwhile to go to Harold's cabins at Snake Lake, so I called up to Harold's and he set us up with the one with the indoor facilities and the screened porch.

But before relatin' the details of our outdoors adventure, and just so you don't get the wrong idea, I'll tell you right off that "Snake Lake" ain't its real name. The wrong idea that I don't want you to get is that I'm tryin' to hide the place from you by givin' it a fake name, as if the lake is such a hot fishin' property that I don't dare let on what it's really called. Truth is, its real name is Dam Eleven, which, I guess, is the kind of name I'd give a place if I didn't want anyone to go look at it anyway.

No, the real reason we call it Snake Lake is that it's full of pickerel, which lots of people call "snakes," and not in a complimentary sense like, well, now that I give it some thought I can't come up with a complimentary reason for calling anything—or any*body* for that matter—a snake, which, now that I give it some more thought, is the whole point. Pickerel ain't your textbook renowned game fish, at least not around here. They're most famous for lookin' more like a snake than a fish, and for bein' a pestiferous bother if you're tryin' to catch a bass. The brother-in-law swears they're actually invertebrates, which is possibly a reach considerin' how bony they are, but it does rhetorically encapsulate the generally dark view of pickerels most people around here take of them.

Of course, pickerel are in lots of lakes, along with bass and bluegills and catfish and perch and suckers and carp and such. And it's true that if you fish at Snake Lake long enough you're bound to catch one of those others. But mostly you'll catch snakes. They pretty much run Snake Lake's ecological melodrama. I don't know why, but somewhere along the way since the lake was bulldozed into being back before the war, when Mother Nature was weavin' the lake's web of life she must've dropped a fairly big stitch and the pickerel just took charge.

I'm pleased to say that this has worked out okay, even on what you might call a socioeconomic level. Most people who patronize Harold's cabins aren't fishermen anyway. They're hammock jockeys who go up there for a weekend of serious swillin' and grillin'. The lake's just scenic backdrop for their little getaway. This has resulted in what you might call a circular if ironic benefit for Joe and me. The lake's only fame is that it's full of snakes so hardly anybody cares about fishing there, which means that Joe and me has it to ourselves, which is worth a lot to us because that kind of isolation from the hordes that infest every other lake in these parts is worth a lot to us even if we do have to fish for snakes on purpose to get it.

Naturally it was dark and rainin' coal scuttles when Joe picked me up, but I kind of like that for a way to start a fishin' trip. It has what you might call a certain meditative ambience, besides further guaranteein' that nobody in their right mind is likely to be goin' fishin' in such weather. There's nothin' like the swish of the wipers and the rattle of the old truck's fan to get you in a good frame of mind for soakin' your butt in a wet boat all day. Doreen had loaded Joe up with donuts and the wife had made sure I had a big sack of cookies for later, and we both had full thermoses, so we were all set for the drive.

We like music on the radio on these outings. It does somethin' nice with the sound of the wipers. My usual favorite Friday-mornin' program, "Baroque Lattes with Liam," faded out pretty quick after we got north of town, so I started huntin' round on the dial for somethin' tolerable, which I couldn't find so I settled for some fluffy jazz, but after a few minutes of lame arpeggios and sappy major sevenths, Joe announced that he'd rather listen to the hog prices for the next four hours than that, so I switched it off and looked around the floor of the truck for somethin' better. Joe's mighty fastidious about keepin' trash in the litter bag hangin' off the glove-compartment button, but he keeps most everything else loose on the floor—tollbooth change, flares, chocolate bars, a couple of your smaller-size firearms, extra socks (no pairs), loose fishhooks, ice scrapers, Doreen's travel nebulizer, and so on—and mixed in was quite a selection of CDs, so I fished out some early Sibelius and slid it into the slot. I can't explain why this should be, but I swear that on a long rainy drive, except for coffee itself, nothin' goes with a couple of Doreen's sinkers like *Kullervo*.

The rain hadn't let up at all when we got to Harold's, so we dawdled over gettin' our stuff moved into the cabin and had a couple lagers and a slow lunch sittin' on the porch lookin' out over what we could see of the lake through the downpour. But after we'd dragged lunch out as long as we tastefully could, and as it showed no sign of stoppin' and we couldn't come up with any other displacement activities, Joe backed the boat down the ramp and we went fishin'.

Forgive me for sayin', but the best way to catch the snakes at Snake Lake is with a fly. Joe figured this out a few years back. If you ever fished for snakes—and I say that in the nicest possible way, knowin' lots of folks might take offense at the very suggestion they'd intentionally do such a thing—then you'll have noticed that most of the time they'll try to eat pretty much anything that isn't actually shootin' at them. So you might wonder what I'm talkin' about, suggestin' that there might be some trick to catchin' them.

But that's another thing that makes Snake Lake so special. A lot of the time you have to work for them, just as if they were a decent, self-respectin' sport fish. About all that Joe and me can figure to explain this is that whatever Mother Nature had in mind by making them the Big Bubba of the lake, it had some kind of strange eco-behavioral side effect, and as the snake generations passed they just got choosier and choosier.

Anyway, one day four or five years ago when we was bein' blanked on snakes again and had to settle for a few pumpkinseeds we'd lucked into, Joe was trailin' a big spinner with a long bit of white bucktail hangin' off it, and as we ran the boat from one spot to another a big snake grabbed it. When Joe got it in he noticed that the spinner had lost all of its parts except for its heavy hook and the bucktail, so next time we were up we took our old fly poles and a bunch of bucktail flies and really cleaned up. Ever since then we have made our Snake Lake trips pretty much fly-fishin' outings and have enjoyed it a lot. Whatever people might say, fly fishing isn't that bad. I even bought a book.

So there we were out on the lake, peekin' out from under the hoods of our slickers as Joe took us a slow circuit along the near shore and I laid a Mickey Finn up against likely rocks and as far into the cattails as I could get away with.

I'd caught three or four little snakes right off, but none of them was worth usin' the pliers on. A couple I even managed to shake off the fly back into the water without even havin' to grab them. I was just startin' to think it was time to offer to switch spots with Joe at the kicker so he could catch a few when I hooked another one that didn't act right. Normally the snakes jump once or twice and make a ruckus on the surface, but a few seconds after I'd hooked this one it sort of sounded and made a beeline past the bow of the boat straight out toward the middle of the lake. More than that, it was so strong that it started to pull the spare line through my fingers and pretty soon it had the whole line and was yanking a fair yardage of backing off the old Pflueger.

Joe and me don't go in for your fancy leaders that run down to a bit of cobweb at the end. The Mickey Finn was just tied onto a long piece of 24-pound mono. I find that a good heavy leader is a big help with clearin' out unnecessary cattail patches on your backcast and your smaller snags on the bottom. Besides, it wouldn't matter to the snakes if I'd tied the fly right onto the end of the fly line. That bein' the case, as this snake chugged away from us, I hauled back real hard on the pole, just to set the hook a little better and let the snake know I wasn't goin' to put up with much more of this. Lo and behold if the snake didn't give at all. As hard as I hauled back, the tip of the pole didn't lift an inch. The rest of the pole just bent more and more. Tactically speaking, this was brand-new territory for us, and line just kept peelin' off the reel, so I suggested to Joe that maybe we'd better follow the fish before I ran out.

He agreed but wouldn't you know it, right then the kicker, which didn't ever run real smooth goin' slow and got fussy if we didn't open her up now and then, chose this as a good time to take a breather. While Joe was tryin' his considerable best to threaten it back into action, I made a strategic decision that if I didn't want to have the reel cleaned off of all its backin' I had better just clamp down with a tight grip on the line against the handle of the pole, which I promptly did.

What with the heavy rain and my slicker blockin' out most of the landmarks along the shore, it took a few seconds for me to realize that we was still followin' the fish anyway, and a second or two after that to realize that it was towin' the boat behind it.

"Would you look at this?" I said back over my shoulder at Joe, who was still busy in the engine room. He looked up, squinted into the rain, and took in the situation a good deal quicker than I had.

"Never heard of a snake doin' that," he said. "Did you see it at all?"

"Yeah, I saw the wake it made as it came in, and then I got a quick look after it hit," I said. "Weren't no bigger than the others."

"Well, my guess is you're not playin' the snake any more. Somethin' else must've grabbed it."

"Sounds right to me," I said, and thought for a minute. "How big a bass would it take to do this? I mean, it's not like we're raisin' a rooster tail or anything, but we're movin' along pretty good."

"Take a pretty hefty bass, I reckon," Joe said. "But it don't matter what it is. What matters now is that we need a good strategy for gettin' it into the boat."

"You're sayin' we need somethin' more, uh, deeply calculated than just hangin' on for dear life?"

"For sure. The way I figure, whatever that is down there, it's either got the fly in its mouth or it don't."

"Depends on which end of the snake it grabbed," I said.

"Right. Even if it grabbed it crossways in the middle, it don't yet have the fly in a position that could hook it. So I suppose, and I offer this just for discussion's sake, if you want to be sure it's hooked, you might be best off to give it a lot of slack so that if it hasn't already swallowed the snake it can go ahead and do that and give the fly a better chance at gettin' a purchase somewhere inside the whatever's gullet."

I gave this some thought and said, "Goes against all the traditions of the sport, don't it? I mean, the whole 'tight lines' paradigm is right out the window at this point, right?"

"Looks like it to me," Joe said, "except that it all depends on how the *snake* is hooked. If he's got the fly down in his gullet, it ain't goin' to be able to catch on anything inside the whatever. No matter how deep it swallows the snake you're liable to just yank it back out of his mouth if you give a good pull." All this time the whatever continued to tow us along steady and the rain beat down so hard that sometimes I couldn't even see where my line disappeared into the water.

"Well, it's a sure thing we can't wait around out here in the middle of the lake for the whatsit to settle down and digest the whole snake so that the hook is exposed and can hang up on something in its gut."

"Granted," Joe said, then added, more brightly, "if the kicker worked I suppose we could try to drag the whatever back to shore. Maybe havin' the snake put up such a fight might rouse its, what you call, your martial instincts and it would keep ahold."

"Well, whatever we do we'd better do it soon. Even if I do drag the whatsit up close to the boat, as soon as it sees us it's just going to let go."

"Granted again," Joe said, "though the rain might help us there. I don't expect the whatever can see through it any better than us."

"On the other hand, it ain't rainin' under the water, and the whatsit will see the bottom of the boat from underneath as soon as we get its head turned toward us," I said, and then added, "Oh, the hell with it. I'm startin' to feel like Santiago here. Why don't I just start pullin' and reelin' until somethin' gives?"

"Like as not it'll be your pole," Joe said.

"And if I *do* happen to get it in close to the boat," I added, ignoring his prediction, "maybe you could see if you could grab ahold of its jaw real good with the pliers? And once you have it I'll drop the pole and try to grab the other end."

"Good a plan as any," Joe said, reachin' for the long-nosed pliers but clearly wonderin' if the plier nose was long enough to keep his hand far enough back from the whatsit's mouth, whatever sort of mouth it might have. "Whoever thought we'd ever need a net on Snake Lake?"

"Not me," I said. "Here goes." Keepin' one hand on the pole handle, I moved the other one about three feet up the pole and hauled back as hard as I could. There was a second there when I swear I heard little crackin' noises from the fiberglass up and down the pole, but then I felt the weight on the line ease up as the whatsit slowed down. Still haulin' hard, I brought my other hand down and started to reel, and to my surprise the fish began to come up.

The Pflueger wasn't made to take line in quick, but after a minute or so of reelin' and gruntin' and pumpin' the rod back like they do in those movies of guys catchin' swordfish, I could tell the fish was gettin' close. I swung the pole round toward Joe, and he leaned the pliers over toward the water a little, just to show he was game for this, and said, "Ain't anybody goin' to say 'this is it?'"

Just then what looked like a big green shovel blade poked through the surface and waved around for a second in the rain before I realized it was the whatsit's tail. It must have been swimmin' a bit against my pull all the way up, so its hind end showed up first.

"Christ in a sidecar!" Joe yelled as the tail went back under the water. "Should I have grabbed that?" The water was so broken up by the rain that we couldn't see into it at all.

And that was it. That was all we got. I was gettin' ready to give one last good haul when the rod whipped up straight and the fly and the front half of my snake came sailin' up out of the water and back over our heads. The whatsit never did let go after all. The snake just finally tore in two.

We sat there in the rain for a minute, and then I reeled in the half snake and suggested that I could use a beer or three with a coffee chaser. Joe probably nodded in agreement, though I couldn't actually tell, what with his slicker and all. Of course, the kicker started right up and we headed back to the cabin. The only thing either of us said after that was when Joe pointed out that now we knew that a snake had a lower pound test than 24.

A few weeks later I ran into Herb Sims, the local fish cop. We go back long enough that I was willin' to take a chance on tellin' him the story, and he didn't seem all that surprised. Not only did he probably believe me, he had an idea of what was goin' on in the lake. He figured that back when the lake was new and just bein' stocked, there could have been a baby northern pike or maybe even a little muskie got mixed in with the snakes at the hatchery, and it had had all these years to itself, just layin' low and livin' off the fat of the lake.

It took me a few days to get used to the idea that anyone ever could have been crazy enough to actually raise snakes in a hatchery in the first place, but the next time I saw Joe I told him about Herb's theory and he said, "Wouldn't surprise me. There's a whole slew of legends goin' clear back to the Middle Ages about pike livin' a real long time." We both thought about this for a minute and then Joe said, "And, you know, maybe it explains there bein' so many snakes in the lake. Maybe the pike can catch all the other kinds of fish easier, or the snakes just don't taste as good or somethin'."

"Yeah, I guess that could be," I said, "or maybe it just feels the same way we do about snakes and doesn't usually bother with them, but when it saw that one thrashin' around it just couldn't resist."

"Like as not," Joe said.

Young Men and the Salmon Trout

MIDSHIPMAN MALCOM CAMERON, A YOUNGER SON OF THE SCOT-
tish borderlands, was at unexpectedly loose ends in Monterey that
summer of 1816. His ship, the HMS *Barbour*, had lurched into
the bay in wretched condition, having encountered an unrelent-
ing series of storms much of the way from Drake Passage. Only
hastily patched together in San Diego, she was going to require
at least a week of attention, possibly much more, from the smiths
and wrights. Captain Murchison, in what seemed to Malcom an
act of uncharacteristic and almost certainly inadvertent generosity,
detailed his youngest midshipman ashore to assist the ship's doctor
and naturalist, a forty-year-old Geordie named David Gresham.

Truth be told, the doctor needed no assistance in Monterey,
and especially none that Malcom could provide, as his energies
were primarily focused on the limited but apparently compelling
opportunities for debauchery provided by the small port and pro-
vincial capital. It was a matter of occasional ribald conversation
among Malcom's shipmates that so far on the trip he had refrained
from participating in the opportunities for serious drinking and
commercial affection provided the sailors at every port. Once or
twice earlier in the trip Gresham had generously invited him along
on such adventures, but Malcom was both religiously high-minded

and terribly shy in this regard, and said so. Gresham didn't press him; it would be a very long voyage and time might change the boy's mind.

Malcom was on the one hand seasoned enough to realize that if he had any real assignment, it was to keep a watch over the doctor so that he didn't hurt himself too badly and could be found if needed, and was on the other hand naive enough *not* to realize that the gruff old captain had long regarded the fifteen-year-old boy as only slightly less worthless than the doctor, and was just as glad to have him out of the way during the serious work of readying the *Barbour* for the next, long leg of its journey to the Orient.

Despite the doctor's easily distracted temperament, Malcom had grown fond of him. Among the books in his father's grand library back home, Malcom had most loved Bewick's *History of British Birds*, which he'd practically memorized on rainy childhood days; and notwithstanding Gresham's tendency to bacchanalian distractions the doctor did have a well-informed and sophisticated passion for natural history. On several previous shore excursions on the west coast of South America and Mexico, Gresham had happily broadened the boy's evident enthusiasm for the natural world and the adventures to be found there. The two quickly developed a productive scientific partnership, collecting a respectable assortment of plant, invertebrate, bird, mammal, and geological specimens, which Gresham patiently taught his young Scottish assistant to prepare and, in many cases, to measure, describe, and even draw with promising results.

But, what with Monterey offering Gresham such irresistible recreational dissolutions, Malcom could see that at least until the doctor's appetites were sated there would be none of the exciting little scientific jaunts that he and the doctor had enjoyed earlier during their voyage. Being an impulsively decisive boy, and having a fair self-assurance that he both could and might as well explore this new landscape on his own, he hired a docile burro and loaded it up with what of their collecting equipment he most enjoyed using: clinometer, plant press, specimen bottles, small-animal traps, and especially the light shotgun that was indispensable for bird collecting.

It must be admitted that Gresham had not yet persuaded Malcom to care about geology—the boy just couldn't understand why the doctor spent so much time chipping chunks off seemingly random ledges of stone when there were so many wondrous living things handy for the pursuit and taking—so the rock hammer was enthusiastically left behind. Malcom topped off the panniers with some fresh (if such a word can be applied to any food that was, itself, almost hard enough to require the rock hammer) ship's biscuit, a bit of local chicken and fruit, and a skin of water. Thus suitably outfitted, he set off early one morning and for no particular reason chose the well-beaten *camino* that connected Monterey with the mission at Carmel, a few miles through the hills to the south.

ORNITHOLOGICAL DISCOVERIES, AND OTHERS

As there had been no time for serious collecting since a brief landing at Banderas Bay, much of what Malcom saw was unfamiliar. He and his burro ambled through the scrub, live oaks, and Monterey pine as through a wonderland, at a naturalist-observer's meditative and highly attuned pace. Now and then he passed a local man or family going along the other way on their own business, but it was a quiet and generally unshared morning, just the sort he loved.

As so often happened on these outings, most of his attention was drawn to the birds, whose flights and songs distracted him at almost every step. Like all naturalists of his day, but even more the young and trigger-happy ones, birds were both objects of wonder and inviting targets for Malcom. Blazing away with the little smoothbore, he soon collected several species. Among them he immediately recognized a nuthatch quite similar but not identical to those in Scotland and a magpie that from a distance also looked identical to the ones back home, but whose unfamiliarly bright yellow bill startled him as he bent to pick up the bird. Other species were less familiar, his favorite among them a California quail (not that he knew the modern names of these birds), with its improbable little topknot; though the vivid oranges of a male Bullock's oriole delighted him as well. He saw two or three Allen's hummingbirds, but never having seen such a creature before, and only seeing these as they darted among the trees at a shaded distance, he imagined them to be some very large flying beetle that would require further investigation on a later hike.

But as he descended the final shady slope down toward the buildings of the Carmel mission, his ornithological preoccupations were abruptly swept from his mind. Coming the other way was an Indian boy who looked to be a year or two older than Malcom. He was clad in the unadorned, coarsely woven clothes of the so-called mission Indians Malcom had been seeing the past few days, but he moved with an almost jaunty gait, and for good reason. In each hand, his fingers through their gills and their tails nearly dragging in the dust, he carried two bright, long fish that Malcom instantly and incorrectly recognized as salmon.

In fact, they were what we now know as steelhead, but like countless other European Americans during the subsequent two centuries, that morning the similarities overwhelmed the differences for Malcom, who had never heard of much less seen any fish that looked so much like a salmon but wasn't. For more than a century they would be known by various names, but often—just by default and for want of any better or more comprehensively descriptive a term—as salmon trout. How could you go wrong with such an all-embracing name?

Not that such nomenclatural specifics mattered that morning, for if there was anything Malcom enjoyed on his father's land more than a long ramble among the parks and forests, it was an even longer day casting for trout on the local streams and on the broad river that bordered the estate. In an instant his thoughts turned from magpies and quail to the possibility that he too might soon be strutting along with his fingers through the gills of just such fish as dangled from this cheerful boy's hands.

His sudden excitement must have shown on his face. The boy smiled widely at him and held all the fish high, a clear invitation to come and take a closer look. Though the two boys had not a word of any language in common they were soon engaged in an enthusiastic exchange of gestures, nods, and exclamations of admiration that must inevitably conclude with the young Scot making it quite clear to the young Ohlone (as his people knew themselves) that he'd dearly love to know where these fish had been caught.

Here it must be said that the reader's first impression of Malcom's new friend, whose name was Tomás, is almost certain to be unfortunately simplistic. Seen strolling cheerfully along with these grand fish, Tomás appears to have it made, but his life story was no Tom Sawyer idyll. Since the arrival of the Spanish missionaries and military forces forty-some years earlier, the native tribes of the California coast were caught up in the often-violent prop wash of international affairs. Generations of Tomás's people and their neighbors endured the usual ugly injustices and unyielding oppression of having their souls saved and their lives redirected into what the fiercely ambitious and heavily armed newcomers regarded as more useful works and ways. Though quite a few of the Spanish missionaries, and even some of the military leaders, had a heartfelt concern for the well-being of their Indian wards, that concern still translated into something that to our eyes today looks an awful lot like bigotry if not slavery. Even on a good day sincerity and good intentions are highly conditional virtues.

That said, on the day when Malcom wandered into view, things weren't going as badly for Tomás and his family as they often had for their native counterparts up and down the coast of the Spanish domain of Alta California. They were at least enjoying all the consolations that their compromised political status and freedom could provide. They were third-generation Christians living comfortably in their modest village near the tranquil little mission church and its kind and generally good-natured *padres*, who largely sheltered them from the worst impulses of the boisterous and randy soldiers a few miles off at the presidio. Like his parents, who had themselves known no other life, Tomás responded enthusiastically to the limited educational opportunities offered by the mission. And the vast array of holy days and other church festivities ensured him almost as much recreational slack time as Malcom had enjoyed growing up in a wildly different cultural and natural setting on the other side of the planet.

Easily catching the drift of Malcom's interest in where such fish were to be found, and waving a fish-laden hand for Malcom to follow him, Tomás backtracked down the hill a short distance. Raising the fish in his right hand, he pointed up the valley of the as-yet-invisible Carmel River east of the dusty road. In a hopeful mixture of Spanish and Ohlone dialect with just a hint of church

Latin thrown in for good luck, he said "There's a really good spot on the river right up that way," which Malcom, though not actually understanding a single word of it, just as easily translated as, "There's a really good spot on the river right up that way." With more nods, gestures, smiles, and a final wave from Malcom and a happy waggle of the fish from Tomás, the two boys parted. Malcom, now almost dragging his burro in excitement, headed toward this new and entirely unanticipated Californian marvel.

There was a good trail to the water, but he missed it completely in his haste, which meant that it was only after twenty minutes of trying to yank the increasingly peeved burro through densely resistant manzanitas that he broke free along a quiet stretch of water that at first glance was not at all what he expected.

What he could see of the river flowed almost sluggishly between narrow dirt banks. It looked no wider than his favorite trout streams, nor as clear. But he heard voices upstream, so he led his burro across the soft bank and around a bend, where his discouragement increased at the sight of several Ohlone men dressed much like Tomás who were just then drawing a long net across a neck of deeper water in midstream. A dozen or so bright, weakly flapping fish were piled on the near shore, and as he watched they were joined by several more, each grabbed roughly from the net and tossed onto the pile.

Malcom was perfectly familiar with the considerable variety of ways in which salmon were harvested, but he stood watching the men work for a moment before he realized that his disappointment, which bordered on outright indignation, was a problem entirely of his own making. In a reflexive burst of homesickness, he had subconsciously presumed that he was about to find well-outfitted sportsmen casting their flies over a river more or less like the ones back home. With this realization, he wondered why he had failed to notice that Tomás was carrying four fish but no fishing gear—no rod, reel, or other kit; not even a handline.

Still, once he'd adjusted to reality and gotten over his embarrassment at the ease with which his enthusiasm had run away with him, he stood watching for a long time. Now and then he could make out the flash of a fish farther up or down the river. Twice, near the far bank of the river back the way he'd come, he saw a few fish work their way through a shallow riffle with their backs exposed to the air for a few seconds before they wriggled upstream into the next pool. A gaudy kingfisher flashed past making its ratchety call, but Malcom barely noticed it, so absorbed had he become in the unexpected joy of being on a salmon river.

Thus preoccupied, it wasn't until Malcom walked on toward the pile of fish that he noticed a young man whose gray robe and flat-brimmed hat identified him as a Franciscan from the mission. The fellow was standing under a tree and watching the river with an absorption very similar to Malcom's own, and didn't seem to notice Malcom until only a few yards separated them.

There must still have been some disappointment showing on Malcom's face. Nothing else could explain the young friar's sudden knowing smile as, without breaking eye contact with Malcom, he nodded toward the busy fishermen and gave a resigned shrug, as if to express precisely the same disappointment. Then, raising his arms so that the sleeves of his robe fell back from his pale forearms, with both hands he pantomimed hauling back on a long heavy fishing rod to which was attached one of the very fish that were flopping sadly on the bank. Then he shrugged again in clear resignation to show that he appreciated and shared Malcom's disappointment at this lost sporting opportunity. Friendships are made in just such moments.

AN ANGLER'S BREVIARY

Like Malcom, Luis Menendez was a very long way from home. Also like Malcom, he was a younger son, but his presence on the Carmel River that day was more a matter of the powerful call of his faith than of the need for a younger son to find some meaningful way to spend his life. His family were merchants in northern Spain who, though not as wealthy as Malcom's, were comfortably well off, enough so that as a boy Luis had accompanied two of his older brothers on several extended business trips on which they combined their business obligations with sampling the trout and salmon rivers that flowed into the Cantabrian Sea along Spain's north coast. In fact, it was on the last of these trips, alone during a quiet morning in an isolated hillside chapel on the banks of one such river, that Luis had found his greater purpose, pledging his life and soul to the church. After considerable training and at the recommendation of his teacher and priest, Luis was chosen for this mission work many thousands of miles away on the unimaginable far shores of the New World.

In miles, Luis's route to Carmel had been much shorter than Malcom's but also much slower. Arriving in Veracruz on Mexico's east coast almost two years previously, he had traveled by cart, horseback, and (mostly) foot, first to Mexico City then to Guadalajara. Both of these were extended stops involving additional training and one long, miserable, and life-threatening bout of "fever" before Luis embarked with several companions on the arduous overland trip north through Sonora to the southernmost missions along the California coast. Once in California, by fortuitous timing he was able to board a small northbound coastal ship at Santa Barbara for the last leg of the journey to Monterey Bay, finally arriving at Carmel as a seasoned traveler, his missionary passion only strengthened by his thousands of miles of hard travel. He was two busy months into his residence there on the day that he encountered Malcom on the banks of the Carmel River.

Malcom responded to Luis's resigned shrug about the fishing with the realization that there was nothing he wanted more right then than to cast a fly to these fish. Being rightly certain that he and Luis shared no common language, he shook his head in disagreement with Luis's fatalistic view, then turned his hands palm up in a mildly supplicating gesture, smiled, and raised his eyebrows, saying in every way but words, "Why not?" Luis's expression was appreciative but skeptical. Malcom, his mind now racing with both the requirements and the possibilities of this sudden and thrilling idea, turned to his burro, rooted around in a pannier, withdrew the quail and the magpie he'd shot that morning. He held them up, spreading and ruffling the magpie's tail feathers as if to say, "We can make *flies*, right? So what's the problem?"

At this Luis's expression abruptly changed from one of doubt to one of sharp interest. From a pocket hidden in a fold in his robe he pulled a small leather bag, from which he extracted his most precious—truth be told, his only—possession, a small but stout book, its cover faded and worn by years of travel and devout use.

The arrival of this little book, professionally handwritten and sturdily bound, on the riverbank is the oddest and most curious part of this tale, and you will not be surprised to learn that the book has a tale of its own.

In that awkward but exciting century following the construction of Herr Gutenberg's printing press, say 1450 to 1550, when similar presses and printers began doing business all over Europe and soon flooded the markets with an abundance of previously unobtainable literary works, there were still a few readers and—probably more important—buyers of books who viewed this spectacular technological leap with grave doubts.

Some of these people, usually those with plenty of money—or with refined taste and the judgmental views that often accompany it—declared that the new and sometimes poorly printed books were nowhere near as handsome as the handwritten ones, lacking as they did the traditional, personal touch that only a one-at-a-time craft production could provide.

Others, again and especially the wealthy or otherwise privileged readers, resented the loss of exclusivity they had enjoyed to that point. In their view, suddenly a great many utterly unworthy people could get ahold of books that had previously only been available to their betters.

Yet others took one look at the new books sitting there in short but shocking stacks and rather than give the matter any reasoned thought, just embraced the fallback reaction of most medieval people to any new thing, proclaiming printers, printing presses, and all their products the work of the devil ("Wouldn't these vile books make a fine pyre upon which to burn their creators?").

And, again more or less predictably, there was one small interest group for whom this stubborn resistance to the new books had the looks of a lucky break: all the scribes whose ancestors had, for generations untold, been the only makers of books. These scribes were in effect the chancers in a trade that was suddenly in big trouble. They were unlike most scribes who, upon getting their first look at one of the freshly printed new books—or, worse, at a stack of them at once—could feel in their guts that their careers were suddenly the historical equivalent of toast. Most of these people, seeing the doom of their craft, took up engraving, illumination, carpentry, tinkering, or any other craft that still showed signs of a viable future.

(It might help to explain here that many illuminators continued to make a comfortable living by providing their usual glorious and colorful ornamental matter to the new printed books. The new bibles and many other texts pouring from the European printing presses still required hand illumination throughout; such intricately colorful illustration was still well beyond the capability of printing presses—indeed, some might say it still is.)

The chancers, on the other hand, whether through good luck, stubbornness, a commitment to their craft, or just cluelessness about what was happening around them, held out. They calmly soldiered on, making books the way they always had. The ones who succeeded in this way did so because they focused on what we now might call niche markets, composed of specifically discriminating elements of the public who, it was to be hoped, would continue to recognize their work as "real bookmaking"—that is to say, books made one at a time, letter by painstakingly handwritten letter, page by page. In some ways, the readers' moods in those restless times were probably not all that different from the crises of identity and perspective we have been living through in recent years as we watch "real" printed books and digital books jostle against one another in the modern book market.

It was one of those hopeful-against-all-odds professional scribes, informally attached to the cathedral in Santiago de Compostella in northern Spain, who in the late 1500s still found a sufficiently sympathetic market for his traditional scribe work. And though he knew that the big money was in unique, large, finely produced, and elegantly illuminated books, he also had some success in maintaining or creating other small markets. Hiring and rather hastily training several assistants, he took a page, so to speak, from the printers' book—exploring every possible shortcut his trainees might exploit to tilt the economics back a bit in his direction, correctly reasoning that even a certain amount of compromised quality wouldn't interfere with his market's desire for something a little more special than the increasingly common printed volumes. After all, in a day when most of your neighbors were impressed nearly to the point of awe by your "library" of twenty-seven books, who was going to be impolite enough to leaf through any of them just to judge their production values?

For the strictly ecclesiastical market, this fellow specialized in small but sturdy compilations of standard texts. Luis's book was one of these. Produced in the late 1600s, it was largely filled with a thorough Breviary, providing its owner with all the essential psalms and other scriptural readings required by the daily offices of the church year. There followed a hefty smattering of this and that ecclesiastical matter, the selection of which probably depended largely upon which texts the scribe happened to have at hand right then but including, possibly on the chance of some salacious market interest, a more or less complete rendering of the Songs of Solomon. It was promptly sold into the Cathedral.

The book that was eventually to come into Luis's hands, thus launched into the cloistered realm, came to rest for long intervals with this or that lower-level church functionary and only moved on to the next at intervals of twenty or thirty years. Finally though, and as these things sometimes happened, it was stolen from the cell of its owner and fenced soon thereafter, thus winding up in a small bookstall in Luis's own town, where his oldest brother, touched by Luis's sudden commitment to the church, noticed it. On a generous whim he bought it for Luis as a loving and encouraging gesture. It instantly became Luis's one great treasure and remained constantly within his reach for the rest of his life.

So it was that by the time the book arrived at the river it was a special object for several reasons. For one, and most important of all, it exemplified, defined, and celebrated the essence of Luis's faith.

For another, it provided Luis with a profound and durable sense of attachment to his family, his village, and his native Spain.

For another, in a century and a half it had accrued a fair though not sensational market value as an object of craft—though nothing compared to the price it would command today should it now resurface in the antiquarian book market.

For yet another reason, and certainly the most urgently meaningful that day at Carmel, its previous owners had felt free to amend its contents in many ways, a not uncommon practice. Most of these additions—all tiny, in at least five different hands, some scribbled and some of fastidious precision—were marginal annotations of one sort or another, intended to elaborate on or facilitate the use of the various sacred texts. Some of these were almost too small to read, others had faded nearly to illegibility. But at some point in its wanderings, the book had fallen into the hands of someone who, in at least one respect, was very much like Malcom and Luis. Which is to say that it became the property of a serious and literate angler.

It has often been observed that most of what we know about the fishing techniques of European sportsmen before the eighteenth century we owe to so few sources, just a handful of books and manuscripts, that we dare not assume that the methods and tackle described in those books were representative of their times. Far more than 99 percent of any generation of these earlier anglers left us no indication whatsoever of how they fished. This is why modern historians greet with such enthusiasm even the slightest previously unknown period-manuscript mention of medieval and early modern fishing. That said, such historians would be beside themselves at the chance of examining Luis's breviary, because the angler who owned it had been both a keen sportsman and an apt chronicler of the means of his sport.

That he intended to create an entire tract on angling, even if he had to squeeze it into the blank spaces in the breviary, is evident from how he started. Imagine him, let's say as an older but reasonably fit fellow in his fifties, his thinning gray hair freshly tonsured, sitting at his candlelit table with the breviary open in front of him, quill and inkpot near at hand. The first thing he does is turn to the final page, the one opposite the inside back cover. Then he spins the book 180 degrees so that what had been the final, blank page of the book is now the opening, blank page of his own new book. And as the breviary's final five sheets were all blank, he has ten "pages" all his own to work with before he must resort to fitting his own neat little script here and there alongside the existing text of the breviary. He stares at his new page 1 for a moment, dips the quill in the inkpot, and begins.

QUESTIONS AND ANSWERS

It has always been a pleasantly absorbing mystery among thoughtful readers of the oldest angling texts, just whom did these authors perceive as their audience? Nowhere is this question as intriguing as in the case of unpublished manuscripts, and we can only wonder to whom this man was writing. Did he have someone, or some group of people—perhaps a brother, either biological or ecclesiastical, or perhaps a younger friend—in mind? Or did he write it all down for less direct reasons, like people today who keep journals for their own reflection and remembrance? Was he purposefully recording his hard-learned wisdom the better to organize his thoughts? Or was he just writing because it seemed like the right thing to do?

Whatever his reasons, he went about the work as systematically as if he had given the matter a good deal of thought before starting. His text was not only written small, it was finely organized and written with the greatest possible economy of expression.

Constrained only by his own priorities, he chose to begin with the construction of his tackle, and to move from smallest to largest, that is, from the fly to the line to the rod. And though Luis had always found the entire text of great interest, it was to the opening three pages, on the making of flies, that he rightly directed the attention of Malcom.

Neither of them had ever made a fly. Malcom's gear back home had all been provided to him by the estate's gamekeeper, and as the boy's enthusiasm was all for getting onto the stream and casting, he had paid little attention to the makeup of the flies except as he tried them until finding one that worked. Luis had likewise just been handed an outfit by an older brother when it was his turn to run a fly through a trout pool or, much less frequently, a salmon run.

Only Luis could read the instructions, of course, but their author had generously illustrated his intentions in a few small but precise drawings that left little to the imagination. Most important, he not only described the making of hooks, he illustrated the result in much finer detail than had the crude woodcuts in the very few printed books of the period.

It's not that fishhooks were hard to come by in the small mission communities; even the *Barbour* had a sizeable stock of hooks among its stock of trade items and for the frequent use of its crew. But these were all large and heavy, not at all suited to the flymaker's craft, and certainly not to the purposes of creating a fly that would satisfy either the anonymous author of the fishing tract or his two young readers. Malcom, the more determined of the two to follow through on this little project (Luis, after all, had formal responsibilities in almost all his waking hours while Malcom could fritter away his time on whatever pleased him), immediately decided that if he was going to do this right his hooks must match the size and shape of the hooks shown in the book.

It was here that the book's author had done them the greatest service. He had not merely described and drawn his hook; so exacting was he that he had laid three different hooks on the page and carefully traced their outlines with his finest quill, thus ensuring that no mistake of misinterpretation was possible.

These precise outlines would be a revelation to many modern fishing historians. The two smaller ones, certainly intended for trout or similar-size species, were to the modern eye of fairly typical proportions, their shanks just a bit short relative to the gape of the bend. If held up against modern sizes, they were roughly a #10 and a #8. The largest of the three was a different matter in its proportions; the nearest modern equivalent would be a #2 hook with a 6XL shank. Each had a barb made in the usual manner, by cutting in at an angle just behind the point and lifting up a bit.

So far so good, but it was not their size or general proportions that would surprise us today, or that in fact so surprised Malcom.

Any angler of his day would have expected the "head" end of the hook shank to be flattened into a horizontal spade shape to provide a bed for the "snell," that is the very end of the horsehair line (our "tippet"). The hair, whether a single strand or many, was firmly lashed onto the hook shank with thread by the fly tier, who then tied the fly over it. But this forgotten angler-author illustrated no such hook. The head ends of the shanks on his hooks featured instead a tight little downward curl of the hook shank that made a small circle, almost closing itself back against the underside of the shank.

This is to say that unlike anything Malcom had ever seen among the anglers on his home streams, these hooks were eyed, in brilliant anticipation of a style of hook-making that would not even begin to catch on, and even then very slowly, among fly fishers in England and America until half a century after Malcom's time.

We're now so accustomed to our fly hooks having eyes that the older snelled hooks we might see in a book of history or a museum look ancient and inefficient to us, though they worked beautifully for countless generations of anglers, including many European and American anglers for a full century after Malcom's visit to Carmel. But considering that essentially all modern fly hooks have eyes—whether down-turned, straight, or upturned—whose circle lies in a plane at 90 degrees to the plane of the hook bend, the hooks in Luis's book would still look a bit weird to us too.

Lacking our conditioning about what an eyed hook was *supposed* to look like, and though never having seen an eyed hook, Malcom immediately recognized the drawing's intent; whatever line he used was going to have to go through that eye and be knotted in place. And of course, knots were something Midshipman Malcom Cameron knew a good bit about.

Raw Materials

For Malcom, this little campaign, at least the part of it involving the making of the necessary gear, was almost entirely about getting the fly right. The other necessities didn't seem an issue. He knew he would need a long rod or pole—the salmon fishers back home typically wielded huge, cumbersome ones of fourteen to eighteen feet—but his morning walk had suggested to him that the landscape would provide a passable selection of tall, straight saplings and stalks that with only a little trimming looked like they would serve.

Having the right line might have perplexed a less easygoing boy; the salmon anglers he knew used finely twisted, knotted, and tapered horsehair lines of laborious construction. Malcom appreciated the helpful physics of a tapered line when making a cast, but the time-consuming handwork of creating such a line (whose preparation was, in fact, also carefully described in Luis's book) seemed not only too much bother but unnecessary. For one thing, the Carmel River wasn't so forbiddingly broad that long, reaching casts were going to be much needed. For another, what with the *Barbour* being totally dependent on the fabrics and control of its sails to get anywhere, the sailmaker's cabinet was always well stocked with a variety of cordage; Malcom could just filch the lengths and diameters required, knot them together, and use as light a terminal section of string as he thought he dared to connect the assembled line to the fly.

A reel (which some Scottish anglers he knew would have called a winch) was still regarded as rather an optional item among his friends and seemed beyond his present resources in any case, so there was no point worrying about it.

He knew that a big fish needed some room to run, though, so he adopted a common expedient of anglers back home. First, he reckoned that he would make up a line of about sixty feet. With a stout pole of fifteen or so feet, he could easily cast about twenty feet of line, giving him a reach of thirty-five feet or so. Then, by attaching a wire loop to the tip of the rod and running the line through that, he'd be able to feed as much more line through it as necessary to give the fish some play should it choose to run. With the blithe optimism of youth, he figured that an extra ten yards of line should be plenty; until he needed that extra line, he'd keep it wrapped around a stick that he either held in a handy pocket or let lie on the ground beside him. No problem there.

It was a matter of a moment's thought for Malcom to think through the preparing of rod and line, though he might need a few days to get it all done. But it was the fly that had to please the fish, and the fly that mattered most.

Even while looking at the drawings in Luis's book, Malcom knew how to get the hooks—plural, because he knew he'd better have several—made. He got along well with the ship's smith and was reasonably sure he could make an arrangement with him to do the bit of work involved in turning out a few hooks, possibly involving in exchange his daily grog ration during some specified period during the next leg of their voyage. A few of the ship's sizeable stock of needles should serve as the raw material. Extracting his journal from a pannier, Malcom made a quick but accurate copy sketch—same size, same curious construction—of the largest of the hooks in Luis's book; only the large hook seemed likely to suit the circumstance with such big fish. He would give his sketch to the smith along with a few purloined needles; he knew that the smith enjoyed finer work of this sort.

But making the fly pattern to tie onto the hook—that was going to take some thought. The book's simple sketches of flies were small, but it was obvious by their proportions which were intended for the long hook; there were only two of them, and they looked quite similar. Both had what appeared to be long, fuzzy bodies wrapped with some fine thread or wire; both had tails half as long as the bodies, most likely made of a small bundle of feather fibers; and both had long wings that lay low over the body and reached to the tips of the tails. Neither showed any sign of wrapped hackles, but the loose fibers of the body's wrapping stuck out along its length in a fashion not all that different from the "legs" created by a wrapped hackle.

The wings were immediately of most interest to Malcom. Like the tails, they seemed to be bundles of fibers—most likely individual barbules—from feathers. Unlike the tails, the bundles were obviously made up of several different kinds of fibers in a style unlike anything Malcom had seen on his trout and salmon flies back home. Even in the sketch it was possible to discern fibers of different shade or color, and at least two fibers with dissimilar but unmistakable crosswise barring.

But the sketch was the only clue. Unlike most other early European writers on fly patterns, who told readers which birds' feathers and mammals' furs to use but gave no hint of how to attach them to the hook or what the finished fly should look like, Luis's anonymous authority did the opposite, with splendid and apparently accurate drawings and nothing else.

Indicating the wing of the flies with a fingertip, Malcom shrugged the obvious question to Luis, who shrugged just the obvious answer back, to which Malcom smiled and silently shrugged, "Right then, what the hell, I'll come up with something." Making quick sketches of the two drawings, he closed his journal, restored it to the pannier, turned back to Luis, and found he was at a loss quite how to say "thank you," much less, "I'll be back soon." But Luis anticipated this problem in etiquette, smiling and making a modest little bow of his head, which Malcom mimicked in return. He was just turning away when he caught himself, again withdrew his journal from the pannier, opened it to a blank page and printed his name. Showing it to Luis, he indicated the obvious question and offered his pencil to Luis, who wrote his own name under Malcom's. This belated formality concluded, the two again smiled and bowed to one another, and Malcom led his burro back to Monterey.

A Fly for Steelhead

It took Malcom four increasingly restless days to get everything ready. He didn't worry about the rod, having scoped out several likely possibilities along the road on his walk home and assuming he'd just chop one down on his way back to the river and rig it up when he got there. He had no trouble cadging the necessary cords and lines, either, or knotting progressively finer pieces of them together in a rough taper that seemed likely enough to carry the fly out across the small river. Having abandoned any pretense of getting everything just right, he figured he would attach the fly to the fine end of this line and not worry about locating some horse-hair for what we would call a tippet. The smith was busy, though, so it took him three days to get around to making up a few hooks to Malcom's curious pattern, as well as turning one needle into a simple circular loop that Malcom could lash onto the tip of the rod.

But the flies were another and more demanding matter because Malcom had fished enough back home—and spent a good deal of time watching others do so much more competently than he—to understand that for entirely mysterious reasons a fly pattern must be "right." As to just what might make his fly right for these fish, all he had to go on was his casual impression that his fellow Scottish anglers seemed to have very little use for bright colors in their flies. That being the extent of his knowledge, he was briefly stumped at how to proceed when Luis's book seemed to suggest the opposite. Because it sounded like more fun, Malcom decided to go with the book. Furthermore, he decided that if color was good, then more color must be better. This meant a discreet visit to Doctor Gresham's cramped cabin on board the ship, where they had stowed the carefully catalogued scientific treasures accumulated on their various collecting trips along the coast of South and Central America. Malcom felt a modest but easily overridden bit of guilt that he was raiding their specimens like this, but the large, silvery fish swimming in his waking dreams obscured his view of any other priority or concern.

Exclusively focused as his attention was on choosing feathers with bright colors and strong markings, he paid little attention to which birds they may have come from, but his rummagings eventually involved a swan, a parrot or two, a glowing hot-pink flamingo, a flying steamer duck, and what would eventually become known as a South American painted snipe. To these he added some finely barred flank feathers from a mallard he'd shot on the way back to Monterey and some of the most vividly iridescent fibers from his new magpie.

He had never even watched a fly being tied, but he did realize that the whole thing had to be held together with thread. His best luck in this regard was finding some reasonably fine red thread while he was rooting around in darker corners of the cordage cabinet, and this was his instant choice for that purpose; more color!

And of course, he had no idea about the dubbing of a fly's body; he figured he would make it of some coarse brown wool yarn he found in the local market.

All these promising materials gathered, and with his small folding knife handy for any trimming, Malcom settled at a table one morning in the snoring doctor's quarters at the presidio and started in. Opening his journal to the page with his copies of the drawings of hooks and flies and picking up the one of the new hooks, the first thing he noticed was the vexing reality that he'd somehow have to hold the hook in one hand—his left, he assumed—while attaching all this stuff to it with the other. But, awkward as that was to prove, he got right to work.

It was dumb luck that he started at the tail end of the hook shank where, right off, the tail seemed to him just the place for some of those bright bits of flamingo feather; indeed, so long were these when he'd pulled them from the bird's skin that he couldn't resist tying in a tail considerably longer than the one shown in the drawings in Luis's book. Fastening the bright red fibers to the hook shank with a few turns of thread, he suddenly realized that he should have got ready the wool yarn for the body in advance, but he managed to raise his right thumb up from where it was holding the bend of the hook and squeeze the tight thread wraps against the shank while he reached over for the yarn, eased it under his thumb with the thread, and then wound it the length of the hook shank before realizing he had no way of holding it tight at the head end of the shank while he wound the thread over it. The fly had looked so nice and tidy—and therefore easily done—in Luis's book, but the execution was proving to be a bit more involved than he expected. Finally he put the head end of the hook in his mouth, clamped his front teeth down on the yarn at the head of the shank, and proceeded to cross-eyedly wind the red thread from the tail along the shank and over the yarn toward his face until he could hold both the thread and the yarn under the fingers of his right hand and then bring both of them into a clumsy—and, for Midshipman Cameron, quite embarrassing—knot, at which point he realized that he'd left no room on the hook shank for the base of the wing, much less the fly's head. But, not having succeeded in making all that tightly wrapped a body, he gave the whole body a few stern twists that not only tightened the yarn and the thread but moved both back toward the tail enough to expose a bit of hook shank near the hook's eye.

For the wing, he gathered a few fibers from each of the several feathers he'd chosen into a bundle, made sure that they were all roughly the same length, and strapped them down on the hook shank as firmly as he could. To his eye, the result wasn't too bad, though the wing seemed a lot heftier than the one in the book's drawing and it tended to lie flatly against the body instead of arcing nicely back over the body as it did in the book. The head seemed about three times the size of the one illustrated, but this was of little concern to Malcom, to whom the idea of a fly having any particular aesthetic "style" had not occurred.

Each subsequent fly was easier for him. About the third try he figured out that he should tie in the tail and body yarn first, then wind the thread up to the head of the shank and tentatively knot it there to await the arrival of the yarn, over which a few turns of the thread would serve to secure it. Then he wound the thread back and forth the length of the body, creating a kind of ribbing that was more involved than in the drawing but looked pretty good to him. After that advancement in his technique, as he tied the rest of the flies he thinned down the wing by using fewer of each kind of fiber.

He then puzzled out how to add that nice little arc to the wing and to getting the head good and tight. In little more than two hours he was admiring the row of six new flies, not only the first he'd ever tied but the first ever purposely tied for steelhead. The sixth and final one especially pleased him.

Fish, On and Off

It was early on Sunday morning when he set off for Carmel with his flies, his line, some tools, and his nearly explosive impatience to finally make some casts. Finding a tall, straight sapling that suited him turned out to be harder than he anticipated; all the ones he thought looked so good the other day now seemed to bend inopportunely halfway up, to be as thick as his forearm around the base, or in some other way not to suit him. When he was almost to the mission he found one straight and slender enough, though no more than fourteen feet long. He was hoping for eighteen. But he cut it down, trimmed off a few straggling branches as smoothly as he could, and tied the metal loop to its fine end before continuing on.

He had occupied much of his waiting-around time the past few days by making a start at learning some Spanish, something he'd resisted doing until now despite all the days spent in the presence of speakers of various incomprehensible Spanish dialects along the *Barbour*'s route. When he arrived at the mission he asked the first passing gray-robed fellow he encountered, in a confounding lowland brogue, "*Donde esta Luis?*" The man, elderly and possibly a little deaf, gave him and his long stick a look mixed of equal parts surprise, annoyance, and amusement, and was apparently considering how to respond when the mission's bell tower came to life with an urgent ringing that reminded the older man that whatever he had been hurrying toward when Malcom waylaid him was still waiting. He spoke quickly and unintelligibly to Malcom, in fact explaining that Luis was surely where he should be on the Sabbath and wasn't to be interrupted.

Malcom could make no sense of this response, but the bells finally reminded him what day it was and it dawned on him that Luis was not likely to be available for some time, maybe even all day, so he bowed politely to the man and turned toward the river. He guiltily realized that he was a bit relieved not to have to share the fishing with Luis, at least this first time.

Now that he knew the path, he reached the river in a few minutes, but the water, so busy a few days ago, was now deserted, and Malcom was dismayed to discover that the fish also seemed to be gone. The riffles where he'd watched fish struggling upstream to deeper pools were quiet, and no fish were evident in the pool where the netting had been so productive. He stood there only a moment before heading upstream; unless the men had caught every single one, there had to be some fish up there somewhere.

He'd proceeded about a hundred yards, for most of which distance the river was shallow and apparently fishless, before coming to a long still stretch that was deep enough that he couldn't quite see the bottom, which was all the encouragement his fading hopes needed. He was so desperate to finally make a few casts that he decided that this uncertain water would do. He moved to the head of the run, looked behind him to see what room there was between him and the brush for his backcast, and set to work. Uncoiling his line and running about fifteen feet of it through the loop at the tip of the rod, he hastily dropped the rest onto the bank next to him. Extracting the sixth fly he'd tied from his satchel, he knotted it professionally to the line and tossed it into the water preparatory to getting a cast into the air; it sank in the shallows in front of him with satisfying quickness and he admired it for a moment as it lay on the bottom in a few inches of water, the long red tail swaying softly in the slow current of the shallows. Like countless anglers before him and since, Malcom's confidence in the fly was half the battle.

Malcom took a good grip with both hands on the butt-end of his makeshift rod, casually palming the line against the "handle" of the rod. A quick flip of the rod moved the fly out into the edge of the deeper current, just far enough for Malcom to make a back-cast, then throw the whole line almost straight out so the fly landed about thirty feet from him in the deepest part of the run and settled immediately out of sight as it began to swing downstream.

He followed this first cast with a couple dozen more, slowly moving down the shore to cover the water as he'd seen fishermen do back home, but had no hint of fish to show for it. He was an experienced enough fisherman not to give up, but he was beginning to wonder if he needed to move farther upstream when his thoughts were interrupted by a most unlikely sound: the giggling of young women. Looking around, he could see no one, but it was clear that there were people nearby, probably in the manzanitas behind him.

Unnerved and embarrassed at the thought that someone was secretly watching him and finding his efforts amusing, he called, "Hello? . . . er . . . Ola?" The giggling abruptly stopped, followed by a rustling in the brush, followed by the emergence of his acquaintance of a few days ago, Tomás, followed by two shyly smiling teenage girls. All three were a bit rumpled. One of the girls was smoothing down her long dress while Tomás was still pulling on his shirt, these actions indicating that maybe my earlier assessment of Tomás's life might have understated just how good he really had it.

But as soon as he saw Malcom, Tomás seemed to forget his companions. He laughed in greeting and in an exchange of talk that this time involved halting verbal communication in addition to vague but enthusiastic noises, the two young men examined and "discussed" Malcom's tackle, especially the fly, whose purpose Tomás instantly and excitedly surmised. The girls, not quite willing to abandon him, stood quietly by as Tomás moved along to the most important subject by indicating the water Malcom was fishing and shaking his head; no fish here! With barely a glance at the girls, one of whom seemed to find Malcom quite interesting, Tomás started upstream and waved for Malcom to follow.

After a quick walk a quarter-mile or so upstream, during which Malcom had to stop once to coil up the extra line he'd been dragging behind him, Tomás stopped well back from a rocky stretch of water where the river had just exited from a long riffle into a deeper run about two hundred feet long and curving slightly toward the north bank. Having no idea that Malcom's primitive fly-fishing gear might obligate him to fish the water in any specific way, Tomás just pointed to the water generally and smiled. He seemed to care not at all that the girls had not come along.

Malcom accepted the invitation. Staying well back from the water like Tomás, he moved to the head of the run, checked the clearance behind him, and landed the fly just where the water poured off the riffle into the run. As it swung around both boys watched the line, and both noticed it jerk to a stop. Malcom swung the rod up, at the same time clamping down on the line to hold it tight against the sapling with a happy this-is-more-like-it yell. But his excitement was instantly replaced by dismay. Whatever he had just hooked raced off downstream as if suddenly remembering a pressing engagement in Queen Charlotte Sound. There was no show, no jump, not even mild disturbance of the river's surface; there was just a second of powerful tension as the rod was pulled down hard, then it snapped back and the broken line shot back over Malcom's head.

Malcom looked over at Tomás, who let out a wasn't-that-great? sort of laugh and waved for Malcom to try again.

This took a few minutes. He had to get another fly out and tie it on the line, which he saw had broken several inches up from the fly. Already beginning to wonder if he'd tied enough flies for these fish, he thought it best to trim that finest segment of line off and fasten the next fly to the next section of somewhat heavier line. He also resolved that this time he would make sure to let the fish run, at least the short distance his extra line would allow, which he had to do on the very next cast, which didn't swing even as far as the first one before some invisible force had hold of it. Unlike the first fish, however, this one stayed in place for a moment, sending headshaking throbs up the line before launching itself into the air close enough to Malcom that some of the spray reached him. Occupied with keeping line ready to feed out through the tip loop if the fish decided to run, Malcom was frozen by this abrupt spectacle of unbridled nature. He again gripped the line for dear life, and the fish rushed off to join its pal north of Vancouver Island. Again the wrenching-down of the rod; again the fly-bereft line whipping loose back over his head; again Tomás happily taking all this in with no apparent regard for Malcom's growing feeling of helplessness.

In fact, it was good that Tomás was having such fun; it kept Malcom excited rather than disheartened, and gave him a hopeful handle on the moment as he extracted yet another fly, cut off another section of line, and tied the fly onto cord so heavy that it barely fit through the hook's eye.

That did the trick. It took a few casts before another fish hit, but as soon as it did Malcom took charge. He beat the fish to the punch, hauling back on the rod as soon as he felt the take, thereby evidently throwing the fish off balance a bit. Then he begrudgingly eased out some line as the fish scurried off to the far bank before making its first jump. Tomás again clapped his hands at the sight of the airborne steelhead, made continuous appreciative remarks as Malcom horsed the fish back toward himself, and applauded some more as the fish came out of the water three times in quick succession directly out from Malcom before settling to the bottom.

At home Malcom had watched some very large salmon being played. There was usually a drawn-out process of give-and-take, with many false hopes dashed when a fish brought near the net found strength for yet another uncontrollable run. So he knew that he was rushing this fish. But the success of those first attempts to control it, especially his ability to snub its movements with the heavy cord, emboldened him to an impatient recklessness. His line could tow a dinghy, the "rod" seemed to have no end of flex to spare, and as long as the hook held he wasn't about to ease up. Checking the clearance of open ground behind him, he held the rod tip high and began to back up. A moment later he dragged a wildly flopping twelve-pound steelhead from the water and continued walking backward until it was far enough from the water to ensure that it couldn't squirm its way back in. With a whoop, Tomás rushed over, slipped a hand into a gill, and lifted the fish for both of them to admire.

Over the next hour as he worked his way down and then back up the length of the run, Malcolm landed three more, and four others were hooked but threw the hook, the last jumping so repeatedly after it was free that both Malcom and Tomás were left staring silently at the water long after the splashes settled. At that point in the proceedings, Malcom looked at his little pile of steelhead, the biggest of which was about fifteen pounds, nodded his head at some inner decision, and held out the rod to his friend. Tomás took it with a joyous "gracias" involving a long string of "muchas, muchas, muchas" capped off with a little bow before turning to the river.

He was at least as apt a caster and angler as Malcom, who envied how Tomás could handle the big rod with just his right hand, leaving his left hand free to manage the extra line. Tomás immediately had a feel for the dynamics of the casting process, and soon discovered that by letting out more line as he cast the heavy fly, he could cover water along the far shore a couple yards beyond where Malcom's casts had reached, where he immediately began hooking fresh fish. By the time a fish managed to run off with their third fly in its mouth (the knot finally unwound itself, not a big surprise considering how many fish tried to eat it), there were five more fish on the pile and even this wonderful, generous piece of water finally seemed to be running out of willing fish. Malcom looked at the fish they'd caught, mentally weighed his share, and wished he'd brought the burro to carry them back.

DREAMS OF CHINA

Malcom would tell the story of that day and of the week that followed for the rest of his long and adventurous life. Two days later he was back on the river with Luis who, though he worked hard to maintain the dignity he was sure his calling required, was thrilled to hook and land a few fish, though it seemed to Malcom that there were fewer taking the fly than last time. Two days after that, he returned alone and his flies attracted only three or four indifferent strikes and hooked no fish. He was just packing up to leave when Tomás arrived with his same two friends, who were more openly friendly with Malcom this time and stood disturbingly near on both sides of him while Tomás made a few casts before hooking a larger fish than either of them had yet seen, but it escaped with the fly after a few frantic minutes.

Before leaving the presidio that morning, Malcom had learned that the *Barbour* was to set sail in two days. He was comforted to be able to imagine that the best of this fishing was over; it made it easier to leave. As he explained this to Tomás as best he could, he was more saddened to leave his new friends than to leave the fishing, and in an entirely unpremeditated gesture, as the four of them got back to the road and were about to go their separate ways, he handed Tomás the rod and the remaining flies. He was able to say both "thank you" and "good luck," and to his surprise and delight Tomás said both back to him.

Before setting out that morning Malcom had also intended to make his way over to the mission to search out Luis and say good-bye to him as well, but now that he'd given all the tackle to Tomás it seemed somehow awkward and unfair not to have something of substance to give to Luis. After all, it was his wondrous book that had made the fishing possible. But Malcom hated leaving friends, so he abruptly and perhaps selfishly decided that one difficult good-bye that day was enough. His ship's departure would be no secret here in Carmel, and Luis would surely understand. That settled, Malcom watched Tomás and the girls walk away until they disappeared down a twisting path through the trees, then he shouldered his bag and headed up the hill toward Monterey.

By the time the presidio came into view, Malcom's mind was already swirling with thoughts of China, which he not inaccurately imagined as a sort of celestial wonderland full of strange people, stranger places, and natural marvels. Primed as he was by his discoveries so far on the voyage, China seemed to him the most exciting possible destination. And—haunted and still unsettled by the dark, expressive, and infinitely promising eyes of Tomás's two young women friends who had stood so warmly close to him that morning—he decided that it was also about time for him to broaden his social horizons in pace with his geographical ones. Fishing didn't enter his mind.

Divergent Perspectives on the Kepler-22b Fly-Fishing Expedition

PERSPECTIVE ONE
Interplanetary Trout Unlimited
News Release
Dateline: Palimpsest Lagoon, Ganymede
April 1, 2212
Contact: Juliana Marbury, PIO

ITU Kepler 22-B Fly-Fishing Expedition An "Awesome Success" Despite Tragedy

ITU's GANYMEDE HEADQUARTERS ARE ABUZZ THIS WEEK WITH the triumphant return of "Team Earth," the first interstellar fly-fishing expedition, just back from their exploration of the fishing opportunities on Kepler-22b, a water planet about six hundred light-years from our solar system. Though their trip was cut short immediately after landing the first fish—affectionately dubbed "Izaak" by team members—ITU executive director LaBranche Hardy immediately pronounced the entire project an "awesome success."

Team Earth leader Mottram Cutchin, freshly revived from twelve years in physiostasis aboard the luxury interstellar sport cruiser *Starfisher*, was effusive in his praise of the potential of "Kaytootooby" (accent on the third syllable), as the boys affectionately refer to the planet they so long dreamed of visiting. "It was sure too bad about those guys on Whalers 4 and 5, but everything else went perfectly. All in all, it was a hoot."

Queried about the transport systems that delivered them to the planet's surface and back home, "Cutch," as he is affectionately known by his innumerable admirers, had nothing but praise. "Hyperspace is a trip. Beaming five whalers from orbit to the surface and back—well, the three of them that came back, anyway—went without a hitch." Asked about the generation he spent in stasis on the outward and homeward journeys, Cutchin's only comment was, "Somehow we need to involve more fiber next time."

According to team reports, all five of the sleek custom proton-drive whalers had just settled to the surface, and team members had barely begun casting, when Cutchin himself hooked Izaak. This was in about forty meters of water, where the bottom was in easy reach of his power-diver shooting head.

"I was using a Darkside Deceiver; you know, the one I developed for the terraformed lunar tailwaters. I'd no more than begun my retrieve, when WHAM! he nailed it. Cleared two of the whalers on the first jump. I'd just tailed him and dumped him into the cargo tank when all hell broke loose. Next thing I knew, this big honkin' fish, must have been a hundred meters long, came up and inhaled Whaler 4. It was an amazing sight, a classic head-and-tail rise that took in the whaler like a mayfly on the Test. Almost flipped all the other boats with the waves from the rise form. We're definitely gonna need heavier tackle next time."

Whaler 5 had likewise been consumed before Cutchin could signal the *Starfisher* to retrieve the remaining boats. "I really hated to call it off after that first fish took so quickly, but some of the guys were a little, well, you know, put off by the other guys dying and all."

Izaak survived physiostasis on the homeward journey in excellent condition, and in fact grew from his modest one-meter size to twelve meters by the time the *Starfisher* reached the Ganymede spaceport. In his brand-new porta-quarium home, he (his actual gender is as yet unclear, but his blue mustache has led team members to assume he is male) will soon begin a special ITU "Star-Trout Tour" in cooperation with several leading outfitters and tackle manufacturers, with stops at all the major outdoor shows at every planet in the solar system.

Rumors that famed fly tier Whit Kelson is secretly developing a half-size whaler imitation for use on the next trip are as yet unconfirmed, though sources close to Kelson say that extraordinary quantities of deer hair have been rushed to his corporate headquarters on the shores of Ithaca Chasma Stillwater, Tethys.

＊＊＊

PERSPECTIVE TWO
United Press Intragalactical
News Release
Dateline: Vermiculatium, with Outer Rim bureau reports
GST 008.2 Galactic Disk Rotation 437
Contact: Darth Adverb, PIO

WAR DECLARED!
Brutal Kidnapping of Vermiculatium Empress Sparks Outer Rim Tinderbox!

Though reports are still sketchy and conflicting, it appears that the entire Cosellian Sector of the Outer Rim has been plunged into a historically unprecedented and rapidly expanding binge of slaughter and planet-scale devastation by the kidnapping of Vermiculatium Empress Blue-Barbel The Elegantly Elongated III, about four Galactic Standard days ago.

The political situation throughout the Outer Rim had been growing increasingly tense recently, but no knowledgeable observer anticipated this bold move of kidnapping the empress, which was carried out by contract mercenaries of unknown species and planet(s) of origin.

Upon determining that the kidnapping had been sponsored by their longtime rivals in the Timurthean Federation, Vermiculatese forces lashed out, their battle fleets swiftly plasma-bombing the Timurthean homeworld and all habitable worlds on about forty of its protectorate planetary systems.

Timurthean Minister of Culture Exquisite Object XI was off-world at a competent accessioning seminar at the time of the attacks, and is thus the only surviving member of the governing council of Timurth. In a statement issued from an undisclosed location earlier today, Mr. Object estimated that the incineration of these planets resulted in the loss of "about three trillion sentient beings plus at least seventy-one presidential candidates and oodles of really nifty works of art. It's going to be a heck of a mess for the fine-arts insurance companies, what with the recent elevation of art prices in the Outer Rim. Our hearts and prayers go out, and all that."

The kidnapping occurred as Empress Blue-Barbel arrived for breakfast in the Imperial Casual-Dining Nook at the Summer Palace Gardens. Observers say that suddenly a perfect replica of a Vermiculatese Nasal Leech—the rarest of delicacies and the empress's favorite snack—raced through the room. The empress, naturally assuming it was a special treat from the Imperial Chef, chased it down and consumed it. Just as her attendants began to cheer the graceful style of the imperial predation, she was brutally whisked away on an invisible cord attached to the bogus nasal leech.

Imperial Guards gave chase but were only in time to capture two of the mercenaries' small surface vessels. The vessel carrying the empress was lofted into orbit, where it was received by a mother ship that immediately took to hyperspace.

The captured mercenaries have been of little help in explaining this strange turn of events. Vermiculatese security staff characterized them as "extremely primitive beings, unarmed except for long sticks of indeterminate function." The only being among them with advanced intellectual capacity was the one they call "Depfynder." His vocabulary is very limited but his analytical skills are spectacular. As he has claimed to have been enslaved by the others, the Vermiculatese government is considering offering him sanctuary.

The mercenaries are being held for trial, followed by ceremonial consumption. A Mark XIV hyperspace tracker drone with full plasma armory is currently searching for their homeworld, which, when found, will be slaggized, then towed into the Perpetual Darkness beyond the Outer Rim.

The identification of the sponsor of the mercenaries was determined from a translation of the script on the hulls of both of the captured vessels. Security staff linguists recognized that "Team

Earth" was simply a clumsy rendering of "Timurth," whose empire has now paid dearly for its ill-considered aggression.

During recent tensions on the Outer Rim, there have emerged ever-more complex treaty entanglements among the three dozen empires that currently make up 90 percent of the inhabited planetary systems in the galaxy. Most observers predict that these standing alliances will inevitably draw virtually every military power in the galaxy into the violence. Media commentators agree that most of the galaxy will be rendered uninhabitable before lunch on Friday. They further predict that this will lead to a collapse in the housing market and subsequent economic turmoil on the galactic stock exchange.

Shupton's Fancy
A Tale of the Fly-Fishing Obsession

The Blue Hart Inn
Friday, June 10, 1996

Dear Al:

What was it that Lord Grey said about the train ride
from London to the Itchen—when you arrive, you step out
"amongst all the long-desired things?" Well, I'm amongst them at
last, with the little River Mistle whispering its way past my window.
After a more harrowing than usual flight, and the obligatory Heath-
row Transdimensonal Crisis (careening out of the lot on the wrong
side of the road, reflexively trying to reach out of the righthand window of one of those wretched dinky rentals to grab some ancestral
gearshift), and two days in Londonium with my publisher's British
toadies, my train was a balm and a blessing, and now I'm in for a
weekend here at the inn. I couldn't see it well in the dark, but Stan
says the water is in fine shape for a good rise in the morning.

All is the same. I recognize every lump in the mattress. Dot
and Stan are fine, fat as ever and just as able in the kitchen (how
they survived their culture's crippling effects on good cuisine I can't

imagine), and (I know you're waiting to hear) Megan eases toward her thirties with sublime grace. My God, what is on the minds of the local swains that they haven't lined up for a try at her? If I were even ten years younger (and even fifty pounds lighter, though any more that's hardly enough to show), I'd fill out an application myself. My kingdom for some prolonged eye contact.

I'm settled in the library, speaking of long-desired things, with an appropriate stout, a good pipe, and no goals until morning, and I find that I now have the leisure to censure you more comprehensively for not coming along. Your excuse of only six hours' notice is all too feeble for me, Al; used to be you'd fish anywhere at a moment's notice.

So here's what you didn't let me explain on the phone. My agent finally talked a satisfactorily fat publisher (Collins-Marstowe, if it means anything to you) into the "Great Dinners of History" (or some similar bilge of a title, "history's greatest porkouts," or whatever) cookbook—my long and brilliant sales record, my erudition in all things culinary, blah blah blah, yadda yadda. I'm to write the coffee-table cookbook to end all such extravaganzas, a veritable hotel of a book, where you can check in and get to know the staff. With a regal advance on the way, I saw no reason to waste a chance at the season over here, so I'm going to hit the stream between stints at the Bodleian, the British Library, and other bibliographically delectable collections. Even London food can't dull my appetite for a good manuscript binge; I do love a library.

Your role, of course, is to react jealously to my accounts of piscatorial triumph and the excesses of Dot's table. I will be merciless.

Yours,
F. Martin

London
The Hampson
Monday, June 13, 1996

Al:

My taunts backfired; it poured rain the entire weekend, and I never even made a cast. Arrived here and waddled out of the train feeling like a beached whale from Dot's magnificent dinners. She gets her flour from some other planet; I can sense a gravitational differential in the loaves, as if they were meant to rest on sturdier tables somewhere far, far away.

I know you don't share my passion for a good literature search, so I'll spare you the high points of my week here, except to rejoice briefly that they've finally accessioned a whole batch of medieval manuscript material that has been languishing in some bibliographical dungeon since "right after the war" (When I asked him "which war?" the heron of a head librarian just drew a bead on me down his nose, sniffed, and continued his admonishments about the criminality of ballpoint pens).

Maybe by reading this old stuff I'll finally discover where Limey cooking went so wrong; it surely had to have been several centuries ago, in order to have declined this far.

Martin

London
Thursday, June 16, 1996

A.:

I'm off to the river again tonight, with better prospects for some sunshine and shadow. At dawn I'm going straight to the Dame's bend for a look. If she's rising, I shall dedicate my broken tippet to you.

Back to my pipe.

M.

London
Monday, 6-20

A.:

Rain again, like sheets the whole weekend. No fishing, no Dame, no rise. But Megan almost smiled at something silly I said during dinner (she was serving again—I don't remember a thing on the plate). I'd swear in court that I saw her lip get ready to turn up, just a bit, like someone who hadn't really almost smiled, but had thought about the idea of almost smiling and subconsciously sent a few electrons down that way, just to test out the circuits.

M.

London
Wednesday, 6-22

Al:

Finally got into the new manuscripts today. Slow going, an endless number of almost illegible little pieces of uneven, cardboardy stuff, each one sandwiched in its own special acid-free folder, served up one at a time by the sour-faced little librarian I think of as the Book Nazi. Doesn't matter if the damn paper only has four words on it; I have to fill out a form, turn it over to the BN, and wait while she knuckles off into the catacombs for a while (probably has a sailor back there) and then carries it out to my little table like she's serving up Charlemagne's testicles on a croissant, where I feel obliged to pretend gratitude (each time, of course) and that such a sacred thing will require great amounts of my time. I pick it up with my little cotton gloves (the fingers are never as long as mine), reread the four words about twenty times, just so the interval is tasteful, then return it to its folder and give it back with another request form, and the BN takes the form from me like she's never seen me before.

I don't really need to do all this. I could pirate a thousand sufficiently picturesque recipes from all the books that have totally exhausted my subject (but not the market) over the past 150 years. But I love being in here, touching such old, real, rare things. Most of them are stupendously dull—"Pyke out the best connes and do hem in a pot of erthe, do therto whyte grece, that he stewe therinne, and lye hem up with hony clarified and with rawe yolkes and with a lytell almaund mylke; and do therinne powdour fort and safroun, and loke that it be yleesshed." Got that? It's more cogent than most (in fact, I'm so used to this stuff that this one sounds familiar, even tasty). These old cooks operated without everything today's prissy food snoots take for granted, all the little comforting amenities like baking temperatures and times, proportions, and measurements. This stuff reminds me of my Aunt Polly, who always responded to requests for her recipes with, "Oh, you know, a mouthful of this, a mouthful of that."

But for all that, I'm the first person, the first even vaguely scholar-like person, to see this stuff since its creation half a millennium ago. That has a lot of value to me. But I can see you're bored already, so I'll let it go. There's hope for clearer weather this weekend.

Marty

The Blue Hart
Saturday night,
June 25

Al:

It was the Dame, and for an instant this morning she was mine. She rose for me, Al, and I left my tiniest BWO no-hackle in her hinge. Just for a gasp, I saw the leader tighten and zing free of the surface film and run straight to the back of her jaw; then the tippet parted and the line noodled back toward me. Shit! of course, but Hallelujah! at the same time! Not a genuine BWO in sight, but something told me it would strike her fancy.

If memory serves, that's seven hookups for me, and five for you. And, of course, twelve escapes for the Dame. Figuring in time invested, my per-hour hookup rate is now .2, and yours is .25, so I'm gaining on you.

Stan made all the appropriately obsequious and congratulatory noises upon my return. Bless him, I think his enthusiasm is genuine, though I can't imagine why after all these years of sucking up to us nearsighted fat foreigners. Megan, of course, said not a word. But I think she gave me more of that sublime whiskey cake than usual, though it may have been my imagination. It certainly gave me "dreams in Technicolor," as my Aunt Verna used to say. I wonder if that's why walruses took to the water—they just got too fat to sleep on the ground without nightmares.

After my triumphant failure with the Dame, knowing I couldn't repeat it in the same weekend, I may return to town tomorrow. I've noticed something interesting in the manuscripts and I'm anxious to get an early start Monday so I can follow it up.

Tight reels, screaming lines, and all of that,

Martin

London
Friday, 7–1-96

Al:

I'm definitely onto something here. I can't quite tell you how I know, but I feel like I've stumbled onto some oddly important little thing that won't let go of my attention.

Last Friday, the umpteenth little fragment of manuscript wasn't quite so little. It was a collection of recipes (it's felonious to call such incomplete things recipes, but you don't care), and faintly marked at the bottom of the last "page" (I'm sure you don't want a digression on folios and such) were what I take to be initials—WW.

Now any medieval British literature geek would right off recognize William Worcester (1415–1482) as the most likely signatory. This was kind of exciting, to actually recognize someone, especially someone like WW, who was my kind of antiquary; loved to save recipes and remedies, all that sort of thing. I even shared my discovery with the heron, who, in a blaze of untoward interest, sighted me in *across* the bridge of his nose with his left eye while raising his right eyebrow (I later tried to do this, in private of course, and couldn't). I had no real use for the information, but it was almost sexually satisfying that no one among the library's cataloguers had identified this little tidbit, while I had.

That happened last week, as I say. Then, just as I was about to give up for the day yesterday, I came upon another signed WW manuscript. This one I took the time to check against a known sample of his writing (it matched, mostly), and it kept me awake all night. It also has entirely sidetracked me from "Great Hog Troughs of the Ancients."

Most of it was as dull as the rest have been. It was a recipe for some kind of ghastly fish chowder (at least that's as close to a generic description as I can come for something so execrably described). But at the end was this little note, in the same hand but a little separate from the rest, like it was an afterthought added the next day. It was almost like an advertisement, Al, the only time so far I've seen one of these old birds tell you where to go to get your raw materials.

Just to make sure you appreciate it, I'm going to modernize it, to spare you the primitive spelling. It ran like this: "And the best fish for this pottage [odd use of this word in this context, but you don't care] is to be got from one A. Shptn, who takes his fish ['fysch,' if you were wondering] on every throw."

What is this, Al? Is it mere promotional hype? This guy "Shptn" is a good bet for fresh fish, in fact he's your best bet because he never fails to catch fish, so his fish are always fresher? Is it an early old-boy system—I'll tell good lies about you if you tell them about me? Or was this guy really good? And what the hell did he "throw?" Bait? Lures? Nets? Flies? *Flies?*

Now you may think this a minor thing to get so worked up about, but I don't suppose you read enough to know better. If you were to read something besides those trendy "White Man's Gravity Sports" magazines, you might understand. Here's what I remembered, as I sat staring at that sentence.

Back about 1980 or 1981, a Belgian linguist named Braek-man published an absolutely charming little book on the *Treatyse*. Among other things, he reported on some previously unidentified manuscripts that are contemporary with, or predate, that book. Now Braekman apparently wasn't all that solid in some of his homework—the Canadian medievalist, Hoffmann, said something about him being "cavalier" in his editorializing. But Braekman had found his way to some amazing new stuff—fly-tying advice, that sort of thing—that predates the 1496 publication of the *Trea-tyse*, which, of course, generations of fishermen (who are hopeless ancestor worshippers) have slobbered over as the fly fisherman's Old Testament.

Despite Braekman's little problems, his findings are great fun. What some of us found most intriguing was his mention of an actual fifteenth-century fisherman, by name. It seemed he found the name of a guy, Anthony Shupton, with some of this fishing manuscript stuff, and figured the chap was a local sport from five hundred years ago. Imagine.

Well, let's see, it was about 1984 or 1985, during my fling with the California crowd, I was over here with the usual abundant wherewithal, so I went to the British Library and tracked the Shupton thing down for myself. Sure enough, the name was real (though the library spelled it Sheepton in their catalog). Besides that, it seemed that the whole thing had been gathered by William Worcester! But as I looked it over, I thought Braekman was kind of reaching to attribute the fishing stuff to Shupton (Hoffmann was right about his carelessness), whose name only appeared later in the manuscript, not even on the same pages as the fishing stuff. That made me doubt that we could count on Shupton as even being a fisherman. It all sounded a little thin, but who could forget the name after the initial kick of such an introduction? Not me.

So it jumped right out at me this time, and it's still got my attention. As I say, I have this feeling about it. Well, it's more than a feeling. It's like when I obsessed about tarpon—remember? It's taken hold of me; I even forgot lunch. That hasn't happened since the yellowtails were hitting that morning off No Name Key.

Well, uplifting as this is, I must go. The train approaches, and I need a weekend just to let this all settle. Maybe Megan will distract me, for all the good it will do us from across the room.

Marty

Sunday, July 15, or 16, or something like that

Al:

Sorry to lose touch, but you'll understand. Well, no, you won't, but you'll listen, and anyway, I need to talk some of this out. Let's see, where did I leave this—I should have backed up the file on this damn thing. I guess I wrote last while waiting for the train to the river.

Well, I didn't go. Or, I went, but I just came right back. I figured I couldn't sit still or fish through a whole weekend, and that maybe somewhere in London I'd find something open, a library or a records center or something, that would let me keep working and thinking about Shupton, so I waited at the station (sent Stan home—he was there right on time to pick me up) until a return train came along, and caught it back to town.

I'm on the great hunt of my life, Al, and I'm on a hot trail. I can't tell you too much. More accurate, I *shouldn't* tell you too much. If this leads where I think it might, I don't want anybody but me to be responsible for what I find. But I'll give you a general outline of what I'm up to. I'm still in London, but I've been out of town a couple times, here and there.

I'm determined to track down this Shupton guy. Having no leads at all on him, though, I started with William Worcester, which led me to certain records collections, which led me to certain villages, and so on, chasing around in church records, county records, archives, estates lists, anything that seemed like a warm scent. Most of it was a waste of time, but really, England five hundred years ago wasn't all that crowded a place, and if someone had some substance, they might get their name written down a few times, maybe even on a tombstone or a deed. Then all that had to happen was that all the subsequent Kings and Queens and City Fathers didn't find some stupid reason to burn the church or move the county seat or pave the graveyard, and there'd I'd have it.

Well, I don't have it yet, Al, but I think I'm getting close. I know, I know, right now you're muttering, "Don't have *what*, fat boy?" Hold on, and I'll get there. In a kind of run-down basement archives, in a more than run-down county building, in a less than desirable piece of real estate not all that far from here, I was looking through some stuff on microfilm (it was the *original* microfilm reader, Al, maybe even a prototype—looked sort of like a mutant composter, and I swear I could hear a squirrel running around a little wheel somewhere inside it). The connection of this place to WW was tenuous, but it seemed he had some pals or distant relatives in the area, but the WW biography has some holes in it, so it was hard to tell, but, as I said before, the scent seemed warm, so there I was, just back from a bilious lunch at the local sty, and grinding through the microfilm between eructations. It wasn't looking good.

Then I come upon an estates list, you know, one of those tallies of some dead guy's property, and I recognize the name—he's an occasional correspondent of WW! So I run through the list, my fingertip going up and down the screen on each sheet, and all at once I have it—the debts. Seems things were, uh, financially slow in 1457, and my subject crossed the river without settling up. Owed half the county, including the transaction I was looking for—one "Aty. Shoptn," owed for two "fschng rds [why did these people hate vowels so much?] and divers dobs." Translation: Shupton sold him some fishing rods and a bunch of flies (they called them dobs, or dubs, as in "dubbing," sometimes). So he was a fly tier. So it could have been a *fly* that allowed him to catch fish at will!

That was good enough, but just a little way down the same list, here was my man Shupton again ("A. Shptn" this time), this time taking a loss on what I think was some kind of building or carpentry work. I not only had a fisherman, I had a dealer in fishing tackle, and I had a tradesman. And that's what I'm working on now.

I guess I ought to check in with Stan—they were expecting me, what, two or three weeks ago now. Maybe you could let them know, huh? I can't stop and go back there now, Al. It's looking too good.

Martin

—◦—

July 30, but I don't know what day of the week, Tuesday, I think

Al:

I can't get the fly out of my mind now. If he really had a fly that always worked, or even worked extraordinarily well, what would it be?

I know one thing for sure: it had to be a wet fly. Whatever Cameron and Heddon and Ledlie and the rest of those upstart revisionist historian giant-killers may have proved about the antiquity of floating flies, it has always been obvious to me that most fish are taken below the surface, and most flies have been fished at least an inch or so below the surface. Shupton was apparently in this for a living, and he wasn't likely to dink around with something as time-wasting as a fly that must float.

I also figure it couldn't be a precise imitation—we have too many of those today, and some of them work well but none of them are miraculous, and I'm assuming (I guess) that this fly was literally never-fail, like my mother's piecrust recipe.

See, too many modern fishermen naively assume that those guys centuries ago were somehow stupid just because they didn't have nylon and graphite; actually, they were probably sharper and more observant than we are. After all, they *lived* on the water, and illiterate isn't the same as unperceptive. But, giving them that much, one of the few things I think we probably *can* (God, I love italics—maybe I was supposed to grow up to be the next Schwiebert) do today is make a much more accurate imitation than they could, right down to the operculate gill thingies, and all that kind of microscope stuff that was beyond the technology or knowledge of the fifteenth century.

So, if it wasn't a precise imitation, it must have been what the Victorians referred to as a "fancy" pattern. The designation didn't necessarily mean frilly or gaudy, though that's what they did to the fancy fly eventually, with all their jungle cocks and silver monkeys and married wings and overdressing; it just meant that the fly's intentions and design were not aimed at anything in particular, but at a pleasing kind of hopefulness of appearance that might work for whatever reason.

So I'm now calling it "Shupton's Fancy," and I figure it's going to be just a little surprising. I kind of expect it to be mostly very ordinary, but with one or two features that really raise the eyebrow at first glance.

But what are they? What haven't we tried, the millions of us over the last five hundred years, furiously tying away generation after generation, adding every possible fiber and hair and gimcrack? What could Shupton have done, on his end of this long long episode of experimentation, that a whole lot of other people wouldn't stumble on later? Was it some strange material? Did he give it some action we haven't thought of? So many questions!

And most of all, would he tell anyone? And even more most of all, will he tell me? Would he be able to resist leaving a message for someone, anyone, letting them in on this wondrous gift he'd found?

Of course, if I didn't think so, I wouldn't be here. I'm betting I understand human nature. I'm betting that no one could keep that secret.

I know what you're thinking, Al. You're thinking I've lost it. I don't even know if the story was true in the first place. Maybe WW was just promoting a friend—I don't even know if this Shupton guy was really able to catch a fish on every cast, much less if he did it with flies. Objectively, I know you could be right. After all, in another one of his texts, WW did describe a mysterious bait ointment that was shown him by "a cunning angler." Maybe, you would say, that was Shupton. Maybe, you would object if you were here, this Shupton, if he was that good a fisherman, just had some kind of goop he put on his bait that attracted lots of fish. Well, you go ahead and think that if you want, but you're not here. I know there's more to it than that; I can just feel it. There's a fly in this, and it's important.

Stay tuned. Maybe I can find out. I hope so. I haven't slept two hours a night for a month, and if I have to eat very many more meals of this British dog poop, I'll bust.

Marty

——

Early August, I think

Al:

I foolishly lost the trail there for a while. Spent almost a week looking around that grubby little thatchpatch of a village where I found the estates list, assuming that if Shupton sold the guy there some stuff, and built him something, he must have lived there somewhere *near* the guy. Finally, I went back to the archives, did some more digging, and found out that the guy had lived for several years about fifty miles away, in another slightly more prosperous village. So I went there, thinking that maybe Shupton had lived in that town too, and hit a sort of pay dirt almost immediately.

It was Shupton's town, all right, and the church and the courthouse and the records were in much better shape, and I found several pieces of paper with his name on them. Seems he was more than a journeyman carpenter, which is what I feared (because that would mean little record). He was something like what we would now call a contractor. Had his own shop, too, where he sold all manner of things (my left nut for ten minutes in that store!). Or maybe I'm reading too much into the records I found. Actually, I suppose it's possible, considering the span of the dates, that the store was owned by a son, or even a nephew or an uncle. But it would make some sense that it was him, so I accept it. Two or three of the more solid little buildings in this village are either his work or have had more modern structures grafted onto his foundation. Isn't that a fun thought?

But for all the intense gratification of tracking him down, I've hit a dead end. Parish registers didn't come in until, what, the 1530s or so, and there just aren't that many other places to look for a fairly unknown person like this. There was no estate list or any such luxury for the guy when he died (there might have been, considering his standing), and as near as I can tell, his line died out or otherwise evaporated in the late 1700s.

So here I am, frustrated. *Piscator interruptus.* Guess I better call Stan and see if I can get back on the river, though all the good hatches are mostly over, I suppose. Maybe the Dame will take a Woolly Bugger, if I could slip it past the rules committee.

I'm calmed down now, if you were worried. It happened kind of abruptly, when I got here, even before I realized I wouldn't find what I was after. Don't know what came over me, but I sure wouldn't want Doctor Dan taking my blood pressure any time in the past month. Whatever you do, don't tell him about this. He thinks I'm spending August at a fat farm.

Martin.

⌒

The next day from my last letter!

Al:

I got it! The scent came back on, and I got it, and it's mine and mine alone!

But I get ahead of myself.

I was set to leave first thing this morning, and I after I mailed your letter yesterday I did call Stan and tell him to expect me tonight. Now that I think of it, there's no reason why I can't still get there. Nothing to hold me here now that I have it.

Anyway, what happened was, after I wrote you yesterday, I decided to indulge in the consolation of a nice heavy dinner, pound down a ploughman's special at the local lard bucket. As it happened, the place was one I'd been to a few times this week, one of the buildings Shupton might have had something to do with constructing. It's one of those dingy little pubs the tourists would find almost quaint if they ever got out this way, as long as they stayed upwind. It was comforting somehow, that even though the search had ultimately proven fruitless, at least I had found this physical connection with the old boy. I wanted one of his magic flies, but I had to settle for his house carpentry.

The barkeep is absolutely antediluvian, a turnip-nosed chap with the diction of a goat, and the place is filthy, but you can tell it's had its times. When I got him talking (I gagged out a compliment about the onions), he wound up and gave me a historical tour (why mentioning the onions set him off on that, I don't know). He claimed the building was six, maybe seven hundred years old, which sounded like at least a 25 percent lie, but practically everything inside is covered with cheap plaster, phony paneling, and plastic signs, so it's hard to tell.

But I'd already been thinking about how old it was. All week, every time I'd go in for lunch, I'd fantasize about taking a crowbar to the whole place: as I levered the last window sash free from its frame (even in my fantasies good luck doesn't come fast), a small "jar of erthe" would roll free from its five-hundred-year hidey hole, drop to the floor, and shatter, revealing among its contents a single, odd-looking fly and a short note on how to tie it.

Well, eventually the barkeep said he'd meant to tell me something but kept forgetting it when I came in. His name's Shipley, by the way—close, but no cigar. Anyway, despite the impression I got from the local archives, it seems there were still Shuptons around until World War II, when two brothers by that name were killed and their wives moved off to London. Then there was only a Shipton woman, who kept chickens for the eggs in the 1950s. The barkeep kept talking about her for a while (as I wondered if these were the same eggs in the jar on the bar) until it dawned on me that he was somehow attached to her, from forty years ago. After a while, as he droned on, I kind of tuned out everything but the rasp of his voice and looked around.

The place, as I said, was a mess, what with several centuries of cheap furnishings having been tacked onto whatever was there at first. The windows were mostly still old (a hundred years or so, I mean) glass, the hand-rolled stuff with all the waves and bubbles, and their frames looked a lot older. The ceiling, one of those dark, high, overbeamed things, was so tarred over with human fumes that I'm sure it must have stopped taking whitewash about Cromwell's time. The bar was obviously modern, as were the stools and all the tables and chairs.

But the back end of the room was simply overwhelmed by a monolithic fireplace, bricked up with some kind of ugly "efficiency" stove piped into it. Its mantelpiece was a huge old log, roughly squared, still covered with the ancient axe scars and irregularities of its making, and about twenty feet long. I'd often admired it, just for its durability, but now, with the droning serenade of the barkeep fading a little as he reminisced softly about the chicken woman, with the dull evening light diffused by the streaky windows, and with the ticky-tacky paneling hiding most of the other parts of the room where Shupton himself might actually have put hand to wood, I flashed on the man himself, or maybe myself as the man, and I saw what he would have to do if he really wanted to tell me about his fly.

He'd want to leave a trace, but he wouldn't want to tell everybody. He'd want to tell someone who was sharp enough, and discreet enough, to figure it out without blabbing it to everybody. Someone who could admire him without wasting a world-class secret. As a practical man, a businessman and a craftsman, he must have known how few things last a long time, and how long his secret might have to wait until someone came along with sense enough to recognize it.

If he decided to leave a clue on paper (and I find no evidence he could even sign his name), where would he put it? The written message is a lousy idea. People didn't own much paper in those days, or think much of caring for it. No, he would look for something more permanent. Something more durable.

Ignoring the barkeep, and without even saying excuse me, I rose from my stool and moved slowly toward the mantel. I was looking at it differently now, not to understand its purpose but to penetrate its other possibilities. I said it was squared, but as I got closer I realized that wasn't quite true. The top was flat, and shelved out from the wall far enough to hold some old mementoes—I remember a boot, and a dust-encased lapwing. But the bottom was rounded, especially where it drove back into the wall right above the fireplace. There was only the weakest light under there, but all it showed was the usual assortment of greasy marks and dents, some probably from drunks raising up under it after messing with the fire, some certainly from the axe that worked it into shape. No intricately carved little images, no recessed compartments. It had been a neat idea, certainly a more deserving fantasy than the jar in the wall, but no dice.

Well, rather than walk back to the bar and hold myself up on a stool, I took a seat in a grimy big fanback chair near one end of the fireplace and redirected my attention to the barkeep, who had never broken stride in his monologue even though I wasn't even pretending to listen there for a while. He was still telling me about his adventures with the chicken woman; you don't want to know.

About midnight, another soul had not been in all evening, and he decided to close up. By then we'd completely stopped communicating. I never did eat. I just sat there trying to generate enough gumption to light a pipe before bed. It seemed that all the weeks of squinting at old paper, all the pathetic, gut-busting food, suddenly caught up with me, and I went into some sort of metaphysical coma—no need to move, no need to think. Just sit and inhale whatever ingrown organisms drifted loose from those appallingly unclean rafters.

I might have sat there all night, for all he cared. In fact, I think he forgot I was even there and closed up without telling me. As he walked out the back way, probably just through reflex he flipped off the light switch, leaving me sitting in the near-dark, except for a streetlight dimly shadowing the room from one window. In my condition, I didn't really mind, but it did stir me enough to decide on the pipe, at last. A few seconds and I had it lit. Then I figured that while I was on a roll and had proved I could still move, I might as well leave.

It was a matter of perfect timing, Al. Shupton couldn't have foreseen it, but if he really wanted someone to know his secret, he must have hoped for it. As I rose from my chair, I realized I was still holding the lighter, so to help me weave through the tables in the dark, I gave it a flick. At that moment, still bent partly over, my eyes happened to be on the mantelpiece, which I was seeing almost end-on and from slightly below. As the old Zippo flared up, light hit that wood from an angle the windows could never provide, and just for an instant my mind registered a faint pattern thrown into unexpected relief. I was halfway across the room before the message of my eyes registered a complaint in my brain and I stopped, realizing I'd just seen something considerably more intentional than adze marks. I turned back and resumed my position at my chair, then held the lighter up.

There in the flickering light was a curving parabola, a thin, very shallow cut (more like a gentle groove than a sharp-edged runnel) in the wood, looking for all the world like the bend of a fishing rod, arcing upside down and then slipping out of sight underneath the rounded bottom of the mantel. Moving to the fireplace and crouching so I could get a look up underneath the mantel, careful to hold the lighter so it would continue to strengthen the shadow, I saw the groove end in a very simple little design, even shallower. Light from any other angle would not have shown it, Al, not even from the fireplace itself. The man was devilishly careful. I thought of rubbing the grime off it for a better look, but I figured that might make it too noticeable, so with a combination of running my fingers along the lines and shifting the lighter all around it to catch the shadows of each part better, I could make it all out well enough.

Well, I don't dare tell you more about the design without risk of telling you too much about the fly it depicted. In truth, though, it only depicted a few key elements. I was wrong, by the way, when I said the fly probably had only one important feature; actually there are three features of the fly that raise the eyebrow, four if you count both wings. I guess you'd have to call them wings, anyway.

Tantalized, are you?

Anyway, it was plain I had to just memorize the thing; no drawings, no rubbings on paper, no evidence but what I remember. That's what I did, then let myself out of the pub and headed to my inn. Back at my room, I dug out my kit, clamped the vise onto the washstand, and got to work. His intentions were so apparent that I was pretty sure what I had to do, but I tried a few variations, just to be sure. Then all I could do was wait until light.

How much thought he must have given that, Al! Imagine how he puzzled over what to do, where to put it, and then how much time he must have spent thinking out its positioning on the mantel.

See, other than the windows and the electric ale signs behind the bar, the only light in the room comes from a single fixture in the middle of the ceiling. It has one grimy little bulb in it now—this isn't the kind of place where people want to be seen, much less seen well—but I could see that the receptacle had just been patched onto some older wooden mount that may have hung a lantern or something, maybe one of those old candleholders. Shupton couldn't have anticipated electricity, but I'll bet he anticipated that the pub would always have fairly unimaginative types in charge—"That's where the lamp always was, so why be movin' it?" Of course, that high light over the room just flattened out the contrast of the visible part of the carving—the arched rod that led you back underneath—it gave it no relief, no chance of shadow, even if someone had cut a new window and added light from another angle. Only when the light came from almost parallel with the mantel, and down low, would you see more.

What a game for him to play, so long before Zippos. I wonder who he thought would find it, maybe with a lit candle, or if he just enjoyed the thought of so many years of the local fishermen getting all liquored up and bragging about how good they were at it, and arguing with each other about this fly or that technique, all the while sitting within pissing distance of the Holy Grail?

But you will want to hear about the flies, won't you? Well, just before what could decently be called dawn, I was already out on the bank of this little local stream, about half a mile south of the village. I'd seen it on the way in, one of those poor misbegotten little creeks that's endured a thousand years of cow manure and worse but still might hide a few fish of some kind. I figured it was a good test. After all, it was probably in worse shape in Shupton's time.

Keeping my eyes peeled (you know how it is over here, I must have been fishing *someone's* water!) I tied the most likely of the flies, a #12 on a 4X tippet and rolled it across this little pool about thirty-five feet, up against a steeper bank on the far side (were History ever to find out about this, She would record that I was using the Orvis 7-4).

It was like a magnet, Al. Remember how the pickerels cut wakes after our bucktails at Grand Lake? Three wakes converged on the fly almost instantly, Al. I swear one of them came thirty feet; how the fish kept from colliding I don't know, but I was hooked to something before the fly had sunk an inch. It turned out to be a big dace, and I caught three or four more, and one crooked little brown, before reeling in and getting out of there. There was something unseemly about it all, even if it wasn't illegal.

I had to finish up this letter on the train, which arrived a couple pages ago, and I want to get it in the mail before the weekend. I think that's tomorrow. Anyway, I'll give you a report in a couple days. Maybe send you a picture of the Dame stretched out on the grass, huh? I've got her now. What am I saying—I've got them *all* now.

M.

The Blue Hart
Sunday, August 14

Al:

It's been like a dream. I caught sixty-four trout Saturday, fifty-nine today. What's the club record for a day, twenty-three? I feel like Babe Ruth. Three of them were more than four pounds, even though the new riverkeeper (you won't like him—snotty young twit) claims there's nothing that big here. They came out of weeds and holes and hides that nobody here even imagines exist. This is better than electroshocking.

They didn't even care about tippet diameter. After a while I trimmed it back to 0X, which hardly fit through the eye of the hook, just so I could horse them in and get them off the hook.

It was trickier than hell handling that many fish and letting them go without being seen. (How did old Shupton keep others from seeing his fly? Were there just fewer fishermen then?) It was good luck, though it didn't seem it when I first heard, that I drew the old granary beat. That put me farthest from the Dame (alas!), but it ensured my privacy. As you know, nobody likes fishing down there because "there aren't many trout." Ha!

I've developed a system for protecting my magic, too. Friday night in my room, I suffered an attack of paranoia about someone discovering the fly. Of course, it wouldn't matter much if Stan or Dot saw it when they were doing up the bed or something because they don't know one fly from another. But all sorts of other possibilities came to mind, so I shaved all the materials off the extra flies, so I would only have one. I keep it in an old plastic 35mm film canister in my fleece-patch pocket. Should I have to, I can get rid of it fast, and it's never out where people can see. Twice, guys came by, once it was even the keeper, and I saw them coming from far enough away to get the fly off the tippet and replace it with a Tups or something. It occurred to me just as I was dozing off last night that if I were to suffer an embolism or something, I could at least try to get that one fly out of the cannister and pitch it into the water where nobody would find it.

I'm still on such a high, I can't imagine going back to town for that dreary book project, but I can't just stay here, either, catching these stupid fish by the barrel. Tomorrow I'll check with Stan about when I can get the Dame's water again. Meantime, maybe I'll do some honest fishing with my other flies, but that doesn't sound very appealing to me.

Marty

P.S. I wonder how big I'd have to tie it for Islamorada?

Saturday, August 20

Al:

This is the first I've been able to write. The gout or some damn thing grabbed me during the night after I wrote, and then I got these vicious gutaches, and I puked for two days. It was hellishly embarrassing, because I couldn't even make it to the can, and Stan and Dot (and once I think even Megan, God forbid) had to clean me up. I was so miserable I barely remember anything before about Thursday, when I was able to listen to the doctor they'd called in from down the road. He gave me this raft about my bad habits, sounded just like Doctor Dan except that he didn't seem to mind the tobacco as much.

So I guess the holiday is over. Stan is checking on my flights, so maybe we can have lunch soon. I figure a couple more days lounging around here won't hurt first, and maybe one more try at the water. Maybe I'll get the Dame after all.

Marty

Tuesday, August 23

Al:

I guess I was sicker than I thought. I'm still weak as a kitten; I just kind of plod over to the pot every now and then, and once a day wobble down to the reading room to pick out a few books to sleep with. I'll write again when I can.

M.

Saturday, August 27

Al:

Still here, but doing much better, thank you. They've taken such wonderful care of me, I hate to leave. Actually, I feel better than I have for years. Had to cut down on the swinishness at table, put the pipe away. I bet I've shed twenty-five pounds, though the first half of it was ejected rather abruptly.

A most uncharacteristic thing happened this morning—Megan actually volunteered a word to me. She'd brought me breakfast, for which I was thanking her excessively when she said (in wild violation of the normal restraint of these people, and looking me straight in the face!), "We're pleased you enjoy it, Mr. Martin— we've never been sure you liked our food."

Perhaps all that close contact when I was semiconscious made her feel free to speak. Of course, I was mortally ashamed of myself. I'm afraid she may have heard me some time when I was ranting about how awful British food is. But now that I think of it, I guess I haven't been making much sense, considering my appetites. I never gave it much thought until she said something. I suppose I enjoyed kicking such an easy target—everyone makes fun of British food, you know. But it must have seemed quite odd to them, what with the way I inhale everything they cook over here. The difference between Taste and taste, I suppose.

Anyway, I do have other things to tell you. This morning I put on a dressing gown and with firm step paraded my new health right down to the reading room. I'd been revisiting the best writers more each day, but today I was able to really dig in, and my, what a wonder it's been.

Skues knew! I'm just sure of it. He found his way to the Fancy, and, true sport that he was, he kept it to himself and stayed with the game despite having an easy way out. He could have let all that quest for imitation go, he could have been even more immortal than he is, simply by writing one short note to *The Field* with a drawing of the Fancy. What a man.

But let me tell you how I worked this out. Think back to the last time someone revealed a secret to you, something they'd been keeping from you for a long time. If you're like me, you immediately picture things that happened—conversations, actions, what-ever—that now, in retrospect, you realize were clues. You suddenly see that the secret holder was telegraphing something to you, either consciously or subconsciously avoiding something. Maybe he suddenly was just a bit too forceful in changing the subject of a conversation, or in some other way walked you down a different topical road than the accustomed one. The change didn't even reg-ister consciously on you, or earn a place in your active memory, but it involuntarily comes to mind all at once, and you smile, and think, "Oh, so that's what was up!"

Well, I've been through something like that, and more than once. In fact, I suspect that others must have known, too. First, here's how I worked it out with Skues. You know he's my favor-ite—all those little notes and essays are perfect toilet reading, and the man was just so damn comfortable to read, an old slipper.

Well, I got to reading him during my waking times about five days ago, and of course there's seven or eight books now where most all of his notes and letters have been gathered up, which is a lot of conversation (sadly, I think it would be impossible to figure this out from very many of the older writers because they didn't leave enough volume of output for it to show up). At first, just grazing here and there in his books, I sensed nothing different. But after a while, the hints were there. The first two or three I barely registered. As I was reading, the Fancy would come to mind unexpectedly, but I would just read on, thinking it was just my own obsession randomly surfacing in my thoughts now and then. But finally I had to see it; he'd get to talking about some fly-tying trick, and I'd recognize its possible connection to the Fancy, and then, just a little too abruptly, he'd veer away and continue in a slightly safer vein.

So that set me to looking for others. I don't know that I found any for sure. I have to wonder if Skues found the Fancy on his own or if someone told him (did he track down Shupton from some completely different direction?).

What's interesting is that I think it's easier to tell if someone *didn't* know. I can tell you for sure that Halford didn't, and as little as we have from Ronalds, I'd bet a month of your pay that he didn't either. The only earlier American who might have known was Gordon; there are a couple of hints in his wonderful notes and letters. But of course he and Skues were pen pals, so it's impossible to guess who told whom. I think if I spent the time with *The Fishing Gazette* and *Forest and Stream* or maybe *The American Angler*, I could get a handle on which other writers from back then were in on it. I wish McBride had written more—there was a first-rate mind in a second-rate business.

Gordon and Skues were so open-minded about fly materials and structure, and put so many years into it, that either or both of them could have just tripped over it one day, though actually seeing the elements of it is only the beginning. You'd have to come to the final product a step at a time, combining several unappealing elements, and that means you have to overcome a whole pile of prejudices we've inherited about proportion and realism and "style."

We fishermen aren't really that smart, you know. Well, we may be smart but we're not especially original. We copy, we modify, but we rarely create, being such slaves to tradition, fashion, and our own predispositions.

But back to the reading room. After I'd realized that most of the older writers didn't leave me enough words to justify a search for clues, I got into that shelf of the really old periodicals (I think I'm the only one to open them since the club inherited them), the ones behind the glass. There's an article I thought I remembered reading about, in the *Sporting Magazine* (out of London, I guess) from 1828, with an illustration of some supposed French flies of the time. I'd never looked it up before, but I remember the article—I read somewhere, probably in *The American Fly Fisher*, that it was all supposed to be a hoax, because the flies were so funny looking and were said by the writer to be for carp. But rereading it (what modern English-language outdoor magazine would publish a whole article in French?), now I don't know. Those flies aren't the Fancy, but I could tell that the guy had a hint, whoever he really was. If you were to look that article up yourself, next time you're here, you'd see what I mean about us being unoriginal. There's something there you will never have seen in any other flies since. It would only take you about 30 percent of the way to the Fancy, so I don't mind telling you this.

What it makes me wonder about, though, is how many guys actually tied the Fancy and didn't know what they had. Think how many of us there have been down through the centuries, all those—how many? millions, I suppose—enthusiastic amateurs grinding out monstrous little patterns for everything from trout to bass to bluefish. Even as surprising as the Fancy is, it could happen that someone just made it by accident.

Say you sat down and felt experimental, so you just tied up all kinds of wild things. We all do that at some point. Some of them we never use. Imagine the star-crossed sap who did that, and then jammed the weird-looking thing into a corner of the box and never used it. Or, even worse, used it one day and broke it off on the first fish, and just figured it was a lucky cast, or, even worse, couldn't remember what it looked like well enough to tie another one!

Skues seems to have been the last. I made a quick run through the most prolific moderns, and though there's one or two Brits who might have had a chance to know, none of the Americans do. Most of the more recent bug-fondlers and techno-empiricists, so busy trying to turn a religion into a science (or a business), don't have a clue, though the more I think about some of his recent stuff, the more I think that if LaFontaine could have relaxed a little bit he might have had a chance at it.

Think of all those intense overachieving young fly fishermen you see out there in their handsome outfits, all of them out to conquer this theoretical world, and they don't even understand what's really waiting at the end of the rainbow.

That's probably just as well. Now that fly fishing has become such an industry (a fortune runs through it . . .), I'd hate for some greedy bastard to get hold of this, patent it, and ruin everything. Imagine the stink. All the state fish and game people would be passing regulations outlawing it, but poachers could clean out a creek in hours. All the lotus-eaters would be cranking out essays and editorials on what's to become of our sport now that there's a way to bypass the uncertainty.

They shouldn't discover it for another reason, Al. Very few of us have the strength of will that Skues must have had, to disallow himself the use of such a gift. I sure as hell don't have that kind of control; I want to gimp out there right now and show the Dame who's boss. Skues was right to keep fishing—to just put the Fancy aside and get on with the struggle—but nobody should even know about this. I mean, it's not just that it makes fishing too easy; if that was the only problem, you *could* just outlaw it, like dynamite. The problem is, it makes the struggle irrelevant.

The damned, ironic thing about fly fishing and fly tying is that the whole goal is to discover something just like the Fancy, but if you do the game's over. It's not a sport any more. You win, but you lose. If you bring the Grail home to Camelot, you're going to be celebrated and canonized, and have a great time at the banquet that night, but what are you going to do with yourself the next morning? Nobody should have this thing.

All at once I'm pooped out, so to speak, and I have to get back to my room. More later.

Marty

⌐◄~

The Blue Hart
September 23, 1994

Mr. Albert Boehringer
24385 Post Mills Road
Grosse Pointe Farms, Michigan

Dear Mr. Boerhinger:

Please forgive my not answering your letter more promptly, but we're still trying to restore a little order here. We were pleased to hear that the services went so well. Seeing Mr. Martin stretched out so horribly there along the river, it was impossible to think of him coming to a peaceful rest on the other side of the world only four days later. He seemed to be doing so well, the doctor was sure there was no harm in him taking a little time on the river.

No need to thank us for caring for the remains and all. Everybody from the embassy and the airport was very kind, and the few other guests we had were quite understanding about the break in our attentions.

We were very fond of Mr. Martin. Sad to say, but I suppose we should be grateful the stroke was so powerful. The coroner thought he could only have suffered a moment before he was gone.

But let me address your questions. First, yes, it's true that he had what must have been the fish you call the Dame on the bank. In fact, she was still on his line, and I supposed at the time that the exertion or excitement is what brought on the attack. We are having her done, of course, in the traditional glass case with a nice brass plate. I think the reading room would be the place for her. He so enjoyed it there.

I must say, I always thought it amusing, your stories of this fish you had named. None of the other regulars ever seemed to see it. But when I saw the look on the face of Lance, the new keeper, when we brought it in, I understood a little better what it meant to you. He tells me a ten-pound, four-ounce fish is simply impossible in this water, that a pike would hardly even get that big. I'm not sure he believes even yet that it really came from our little Mistle.

The fly is another matter. You know I don't fish, Mr. Boehringer, but I am obliged to know enough to keep an eye out for the unusual. You Americans have such different ideas about what is proper, but had I seen the fly he was using under less tragic circumstances I should have had to report it to the membership committee. It was one of those enormous hairy things, I think you call them Woolly Boogers? Of course, I promptly disposed of it, and there is no use in saying anything now. I'll say he was using a Tups, when we do the plate.

I believe that takes care of it. Again, please accept our most heartfelt sympathies for the loss of your friend.

Sincerely,
Stanley Frampton
Prop., The Blue Hart Inn

P.S. Rereading your letter I discovered that I had overlooked one of your questions. You are correct that Mr. Martin's vest was somehow left behind; I just found it hanging in the equipment room, and I will have it sent on. At your suggestion, I checked the pocket under the fleece patch, and indeed there was a film canister in it. But it wasn't empty, as you supposed. There was one fly in it. I assumed that you might want it or you wouldn't have asked, so I am attaching it to the bottom of this letter.

Acknowledgments

My thanks first and always to my spouse and partner-in-all-things, Marsha Karle, whose presence and encouragement made the book possible.

At Stackpole Books, I am as always grateful for and in awe of Judith Schnell's publishing vision, from which about a dozen of my books have benefited greatly. I also appreciate Christine Fahey's able management of the publishing process, and Cheryl Brubaker's thorough and thoughtful editing of the manuscript.

At various times Bob DeMott, Andrew Herd, Richard Hoffmann, Patricia Junker, Marsha Karle, Nick Lyons, William Palmer, and Steve Schullery all provided thoughtful and helpful readings of various chapters. They invariably improved my stories, and any flaws remaining are probably the result of my not listening harder to their advice.

Some of these chapters appeared, usually in quite a different form and in some cases rather long ago, in the following publications.

"Second Season" originally appeared in *Trout* magazine, thanks to then-editor Tom Pero. "Selective Chubs" originally appeared as "Chub Water" in *The Flyfisher*, thanks to then-editor Dennis Bitton. It was brought on by a hilarious column by Nick Lyons in *Fly Fisherman* magazine in which Nick outlined the book about chubs that he and Craig Woods were then supposedly working on. "Divergent Perspectives on the Kepler-22b Fly-Fishing Expedition" previously appeared on Midcurrent.com; thanks to editor Marshall Cutchin both for suggesting the story and for running it. "Shupton's Fancy: A Tale of the Fly-Fishing Obsession" was originally published as a book by Stackpole Books, thanks to Judith Schnell and David Detweiler.

The other chapters have not been previously published, though I do have this nagging feeling that "Holocene Outfitters" also appeared somewhere previously and I just lost track of it. In any case, I wish again to acknowledge its inspiration, Ray Bradbury's provocative story "A Sound of Thunder," which first appeared in *Collier's*, June 28, 1952.